Commercial Chicken Egg Production

The Author

Dr M. Murugan M.V.Sc., Ph.D, MBA (Human Resources) is currently working as Professor of Poultry Science in Tamil Nadu Veterinary and Animal Sciences University, Tamil Nadu. He is a prolific writer in the domain poultry science and has inherent passion for teaching. He is also a resource person for human resource management in poultry sector. He has got twenty year of teaching, research and extension service experience in the field of poultry science with special reference to commercial chicken production for egg and meat and Japanese quail production. He has published twenty research articles on poultry science subject in various international and national journals. He has authored a book titled Commercial broiler chicken production. Currently his areas of interest are content development, ICT applications and human resource development through capacity building in poultry science and production sector.

Commercial Chicken Egg Production

M. Murugan

2020
Associated Publishing Company®
A Division of
Astral International Pvt. Ltd.
New Delhi – 110 002

ISBN 9789390371983 (Int. Edition)

Published by : **Associated Publishing Company®**
A Division of
Astral International Pvt. Ltd.
– ISO 9001:2015 Certified Company –
4736/23, Ansari Road, Darya Ganj
New Delhi-110 002
Ph. 011-43549197, 23278134
E-mail: info@astralint.com
Website: www.astralint.com

Digitally Printed at : **Replika Press Pvt. Ltd.**

Preface

The chicken egg is a unique, natural, unadulterated and nutrient rich food for human and animals. Basically, the purpose of egg laying of chicken is to nourish their young ones during their embryonic development and not for human use. However, the history of use of chicken egg as human food is traditional and time immemorial. The history of commercial chicken egg production is recent, when compared to domestication of chicken. Indian commercial chicken egg production system is barely fifty years old. Third five plan period is the starting point for commercial chicken egg production through import of hybrid layers from developed countries.

Development of commercial chicken egg production is a success story of agriculture and allied sectors of India. India stands on third position in the world egg production. Geo spatially, India's rich experience in tropical chicken egg production is unique in nature. It cannot be compared with developed countries production system in terms of rearing methods but, production standards of commercial chicken egg production of India is on par with international standards. The contributing factors of development are elite genetic stocks, scientific production, vibrant marketing network and acceptability of the consumers. Now a days egg is becoming a regular food item in non-vegetarian population especially children.

Standard text books on commercial chicken egg production with Indian farming perspectives are not available for poultry science students. Considering that, the objective of the book is to elaborate the commercial chicken egg production with Indian perspectives and farming systems. The book will be useful for student of veterinary and animal science, poultry science, farm managers and farming aspirants. In this book, the contents is sequentially arranged and explained with examples for easy understanding and activities have been provided at the end of chapter for practice.

Finally, I want to express my sincere thanks to my wife *Abbiramy* and my daughters *Muthamil* and *Manicka* for their moral and emotional support to write this book successfully. I also express my sincere thanks to my Professor R. Prabakaran, who always inspire me through his thoughtful and untiring acting and writing in the field of poultry science.

Dr M. Murugan

Contents

Preface v

1. Introduction to Commercial Chicken Egg Production 1
2. Genetic Stocks of Chicken Egg Production 5
3. Poultry Housing of Layer Chicken 9
4. Layer Chicken Management 21
5. Commercial Layer Chicken Feeding 33
6. Chicken Egg Quality 55
7. Bio-security in Commercial Layer Farm 73
8. Common Poultry Diseases 81
9. Water Quality Management 91
10. Good Husbandry Practices in Commercial Layer Farming 93
11. Commercial Chicken Layer Farm Economics 99

Annexure 103

Index 105

1

Introduction to Commercial Chicken Egg Production

Learning Objectives

1. Know the history of commercial poultry production in India
2. Describe the concepts of commercial chicken egg production in India
3. List the factors responsible for the development of chicken egg production

Commercial chicken egg production is one of the important segments in commercial poultry production. Poultry production deals with the process of rearing poultry species of birds for commercial and economic uses, especially for production of chicken egg and meat. The term, *Poultry* refers the domesticated species of birds, used in commercial rearing for food and economic purposes. It includes chicken (*Gallus domesticus*), duck (*Anas platyrhynchos*), goose (*Anser anser*), turkey (*Meleagris gallopavo*), Japanese quail (*Coturnix coturnix japonica*), guinea fowl (*Numida meleagris*), *ratites* such as emu (*Dromaius novaehollandiae*) and ostrich (*Struthio camelus*). The study of these poultry species is called as *Poultry Science* and, the study of all the birds is collectively called as *Ornithology*.

Documentary evidences of domestication of poultry suggest that, Red jungle fowl (*Gallus gallus*) was domesticated as early as 3200 BC. Egyptian and Chinese records indicates that, laying of eggs by fowl for human use was dates back to 1500 BC. Chinese records described fowl as domestic time clock, due to cock's early morning call and hen's regular egg production. The number of breeds and varieties of chickens evolved over the years in worldwide perspective is almost around two hundred for egg and meat, but only a few are economically important as egg producers. Commercial chicken egg production is the process of rearing commercial layer chicken for egg of food purpose. The commercial layer chicken is a hybrid chicken genetic stock, selected, bred and reared exclusively for table or infertile egg production for food consumption. Most laying hens in the leading chicken egg producing countries are improved strains of Single-Comb White Leghorns for white shell egg and Rhode Island Red, New Hampshire and Plymouth Rock for brown shell egg production.

The commercial poultry production industry of India is around fifty years old and, considering the history of domestication of chicken (around 2000 BC); it is very recent in origin. In India, the Christian missionaries were the first in the start of commercial poultry farming in early part of 20thcentury. The third and fourth five year plan (1961-1974) were the notable periods for the development of poultry industry of this country. Intensive Poultry Development Programme was introduced in Third Five Year Plan to provide all essential inputs for commercial poultry farming. Central government institutions like Central Poultry Development Organizations (CPDO), All India Coordinated Research Projects on chicken egg of Project Directorate on Poultry (PDP- Currently Directorate of Poultry Research – ICAR-DPR) of Indian Council of Agricultural Research (ICAR), New Delhi, Central Avian Research Institute (CARI), Bareilly, Uttar Pradesh, and state agricultural and veterinary universities were the key organizations involved in popularizing, promoting poultry industry in India. The role of private players or breeding companies in poultry production started in early seventies by importing improved genetic stocks from developed countries and establishing own pure line breeding companies in India. A number of hybrid chicken stocks under various brand names are available in India now for commercial layer chicken farming through these private breeding companies.

Currently, commercial chicken egg production industry is highly scientific and industrial in nature. The production system is characterized by *high input - high output* production mode in its operation and efficiency. Commercial chicken layers are reared up to 72 weeks of age and above, producing a flock average egg production of 300-320 eggs per year of egg production. Intensive system of rearing is practiced with cage rearing methods and high level of automation in management. In commercial rearing, people preference is for white shell eggs in India.

The leading egg producing countries of the world are China, United States of America, India, Japan, Mexico, and Brazil. India stands third in egg production after China and USA and producing 72 billion eggs per annum with the per capita consumption of 64 eggs per year. The leading egg producing states in India are Tamil Nadu, Andhra Pradesh, Telangana, Karnataka and Maharashtra in South, Haryana and Punjab in North India. As per Indian Council of Medical Research (ICMR), the recommended per capita consumption of egg for human is 180 numbers per annum (half an egg per day). Comparing the recommendation, availability is around one third only. The year wise egg production and per capita consumption of eggs in India is presented in Table 1.1.

Table 1.1 Egg production and per capita consumption in India

Year	Egg production (billion numbers)	Per capita consumption (in numbers)
1950	1.83	5
1960	2.88	7
1970	5.30	10
1980	10.06	15
1990	21.10	25
2000	30.44	30
2010	63.02	53
2013	69.73	58
2014	72.50	64

Source: FAO statistics on egg production - 2014

Critical Factors for Development of Commercial Chicken Egg Production

There are number of factors, which support the development of commercial chicken egg production in India and they are:

1. People's preference over egg as a quality and nutritious non-vegetarian food
2. Good source of animal protein comparatively at minimal cost
3. No religious sentiments or taboos on egg consumption
4. Improved income and purchase potential of people and rapid urbanization
5. Import and development of egg type pure lines chicken for commercial use
6. Scientific rearing and management model

Now a days, commercial chicken layer industry is become more industrial oriented, which needs substantial capital investment. Though, the development commercial chicken egg production system is high in the recent past, the industry is also facing some serious problems like increase in the effective economic size, substantial capital investment, escalating feed costs, food safety, chicken welfare and pollution issues etc. However, understanding and implementation of *Good Husbandry Practices* (GHP) in commercial chicken egg production will be useful to overcome these issues by larger extent and to improve the marketability of the eggs and consumer satisfaction.

Activity

1. Compare the egg production data of various states of India
2. Compare the data of leading countries in egg production

2

Genetic Stocks of Chicken Egg Production

Learning Objectives

1. Discuss the concepts of commercial chicken layer breeding
2. List the critical traits of selection in commercial layer chicken breeders
3. Discuss the generalized industrial breeding protocol in layer breeding

Genetic stocks of commercial chicken egg production are essentially *hybrids*, which means, the stocks are expressing *hybrid vigour* in their production performances. The hybrid vigour of these birds is achieved through systematic selection and various methods of cross breeding programme for considerably a longer period. These genetic stocks have been selected and bred for high egg number and other related traits of egg production. In case of commercial layers, the classical breed of *Single comb white leghorn* breed of chicken of *Mediterranean* class was exclusively used in primary breeding to develop commercial strains. The advantages of this breed are small body size, white shell eggs, egg number and feed efficiency. Since, majority of the world prefers white shell eggs; this single comb white Leghorn breed of chicken is considered as a universal choice and dominates the commercial chicken egg production markets all around the world. However, in some of the areas, brown shell eggs are preferred, for which the classical breeds of *Rhode Island Red, Barred Plymouth Rock* and *New Hampshire* of American class are used in primary breeding process to develop brown egg shell laying strains. The commercial chicken layers are presented in Figure 2.1.

Layer Chicken Breeding

Initially, the classical plant breeding principles of hybridization of corn were used for the development of pure lines for sire and dam. Nowadays, almost all the commercial strains are *inbred hybrids* (crosses of inbred lines) of almost a closed population of chicken stocks. Currently, the role of government organizations in pure line breeding of chicken stocks is very minimal in India and they are in the hands of private and corporate breeding companies. Every primary breeding company is having their own commercial strain in a brand name, supplying grandparent, parent and commercial stocks along with specific management recommendations for profitable rearing and performance.

Figure 2.1. Commercial chicken layers

White egg shell layer

Brown egg shell layer

Source: http://www.hyline.com

Selection objective of layer chicken breeding is comparatively straight, when compared to broiler chicken breeding. However, the challenge is improvement of egg production in terms of egg number that is the number of eggs produced in a given laying period. The egg production is basically a *fitness trait* with very low heritability, will not respond well for conventional individual or mass selection methods. Moreover egg production is also a *sex limited trait*; thereby improvement in egg production is achieved through individual, pedigree, family and combined selection tools and by constructing complex selection index figures for overall estimation of genetic merit of the individual. Mostly multi trait selection index figures are used to make the selection more efficient and wholesome. Chiefly, selection index scores includes the critical traits like age at sexual maturity, body weight at sexual maturity, egg number, egg weight, feed efficiency, shell thickness etc. In layer chicken breeding, except egg weight all other traits are having low to moderate heritability, makes the selection progamme complex. The major and minor traits of focus in industrial layer chicken breeding are presented in Table.2.1.

Table 2.1 Selection traits in industrial layer chicken breeding

Major traits	Minor traits
Age at sexual maturity	Disease resistance
Body weight at sexual maturity	Heat tolerance
Egg number	Pullet eggs
Egg weight	Egg abnormality
Feed efficiency	Blood and meat spots
Egg shell thickness	
Livability	
Fertility and hatchability	

Generalized theoretical industrial breeding plan of layer chicken has four levels namely, Pure lines, great grandparent, grandparent, parent and commercial stocks. Here, pureline, great grandparent, grandparent are purebred in nature and parent and commercial are cross products. The primary breeding company will supply only unisex chicken stock to grandparent breeder growers to have a security check on replication of grandparent for further use. The graphical representation of the generalized industrial layer chicken breeding programme is presented in Figure 2.2.

Figure 2.2 Generalized industrial layer chicken breeding programme

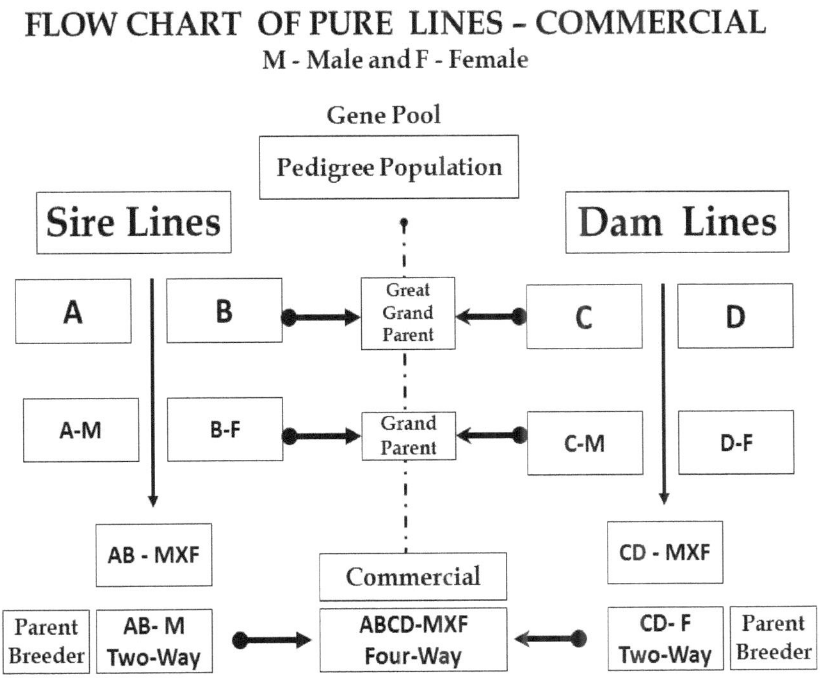

The various steps involved in the breeding programme are as follows:

1. Pure line breeding with pedigreed population
2. Formation of inbred lines (exclusive sire and dam lines)
3. Testing of general and specific combining ability between lines
4. Selection of lines with good combining ability
5. Crossing of inbred lines to exploit *heterosis* or hybrid vigor
6. Supply of unisex (male or female) progeny at grandparent level
7. Crossing of four lines of grandparent to develop parent breeders
8. Crossing of two parent lines to form commercial bird

Based on the above programme, it is evident that parent and commercial layer chicken are two way and four way crosses, which is bestowed with hybrid vigour due to the inbred line crosses in the selection and breeding history of its formation. The purpose of making parent and commercial laying stocks as two way or four way cross is to exploit the heterotic effect on egg production, which is essentially a fitness trait with low heritability.

The student of poultry science should understand that, whatever the improvement in the performance is achieved in commercial layer chicken stocks are only through conventional and advanced selection, pure and cross breeding methods and not through genetic engineering or genetic modification. Based on that, commercial chicken layer stocks and its eggs are not *Genetically Modified* (GM) or *Genetically Engineered* products as most of public thinking. In addition to that, recent researches in developed countries clearly demonstrates that, though use of genetically modified feed ingredients like Maize (GM) in poultry feed, it does not pass on any of its foot print to the eggs.

Understanding the superiority of these commercial chicken stocks is highly essential to plan and execute the management programme. These birds have to be supported by good quality housing, feeding and health management progamme; otherwise desirable results cannot be achieved. The major layer chicken strains available in India are BV 300 (white egg shell) BV 380 (brown egg shell), Hyline white, Hyline brown, Bovans, Lohmann etc. It is always better to know that, individual primary breeding companies recommend their own management and feeding standards for their commercial strains, in which mild variation exists between commercial strain performances and their management specifications. However, understanding the basics of this variation will help the practioners to devise specific plan for a commercial strain in question.

Activity

1. Compare the production performance of various commercial layer chicken strains available in India and offer your comments.
2. List the critical traits in layer chicken breeding with their heritability and offer your comments.

3

Poultry Housing of Layer Chicken

Learning Objectives

- Discuss about the rearing systems of commercial chicken layers
- Discuss the cage specifications for commercial layer chicken
- Design an open sided elevated cage house for commercial layers
- Discuss the concepts of enriched colony cages for chicken layers

Introduction

In commercial layer chicken farm, chicken houses in the form of brooder, grower and layer houses constitutes nearly 50-60 per cent of the total capital expenditure, depending on the type of house. The purpose of housing is multifactorial and they are:

1. To protect the birds from extreme climatic conditions, like hot sun shine, rain, cold, theft and predatory animals.
2. To provide comfortable micro climatic environments and rearing conditions for better management to get desirable performance in production.
3. To facilitate automation in large farm operations like rearing, feeding, watering and egg collection.

Systems of Rearing

Basically there are three systems of rearing in chicken egg production and they are:

- ***Extensive system of rearing:*** It is the traditional system of rearing, otherwise called as *free range system* of rearing, in which, there will be provision of night shelter only for birds. Grazing is the main source of feeding along with some supplementary feeding. It operates on the low input and low output production mode. It is suitable for backyard or family based rural chicken egg production. Birds of indigenous genetic stocks are suitable for this system.
- ***Semi intensive system of rearing:*** In this system, a small scale shelter in the form of chicken house will be provided for the birds. Grazing is the partial source of feeding along with regular supplementary feeding of concentrates or compounded feed. It is suitable for small scale country chicken egg production in deep litter rearing and operates on medium input - output mode.
- ***Intensive system of rearing:*** In this system, well designed and structured houses will be provided for birds. Systematic feeding and watering will be made available in the house itself. This systems works on high input - output production mode. This is suitable for commercial or industrial chicken egg production.

Commercial chicken production system for egg is characterized by intensive system of rearing. In this system, birds are reared in well-designed chicken houses with reasonably comfortable micro climatic environments to achieve good production potential. Environment within the chicken house in terms of temperature, relative humidity, ventilation and air quality is called as *micro-environment*. Desirable micro-environment for commercial layer chicken is presented in Table 3.1. However the climatic profile of India is diverse in nature. The ambient temperature during summer may high, in the range of 35-42° C (95 to 107.6°F) and in winter it may go as low as 7-8°C (45 to 46°F) in central lands. The relative humidity in coastal region may be around 80-90 per cent and in central lands, around 40-50 percent. Considering the temperature and relative humidity, the area may be classified as hot - dry, hot - humid, cold - dry and cold - humid with a mark of 25°C and above as hot and 50 per cent and above as humid. Hot and humid is the most stressful conditions for animal farming including poultry. Moreover, birds do not have sweat glands; make it more complex to poultry farming in hot and humid areas. Considering this issue, poultry housing should be a well-designed structure to provide optimum micro environment to the birds for their superior performance.

Table 3.1 Desirable micro environment for commercial layer chicken

1.	Temperature	22-26°C
2.	Relative Humidity	40-60 percent
3.	Ammonia level in poultry house	Less than 20 ppm
4.	Deep litter moisture	15-25 percent
5.	Carbon di-oxide in poultry house	Less than 0.5 per cent

Identification and Selection of Farm Site

Identification and selection of ideal farm site is a serious issue and it involves a number of activities and careful application of mind. In this process, selection of wrong site will be a permanent problem and it will be a pre disposing factor for other managemental problems. The critical factors to be considered for proper site selection are as follows:

Topography: The farm site soil should be virgin to poultry farm operations with no history of previous exposure to poultry farm activities. The farm site should be elevated with good grip soil for buildings, free from water logging and flooding with reasonable wind flow and drain.

Locality: It should be located reasonably near to the market to procure the inputs and market the outputs, and away from other poultry farms and residential areas to avoid future pollution problems and complaints.

Accessibility: Access to transport logistics like road for the smooth flow of inputs like chicks, feed and utilities and outputs like egg, culled live birds and manure is advantageous.

Water availability: The land should have sufficient water sources and water should be free from all impurities to provide clean and safe drinking water to birds.

Power supply: The land should have access to electricity power to carry out the, brooding and machinery operations of a farm.

Labour availability: Availability of cost effective labour is a desirable factor.

Approval: The land should be an approved land for poultry farm operations. In areas, where local rules and regulations exist for the same, necessary permission should be obtained from local authorities.

How to Decide the Farm Area Requirement?

The commercial layer chicken farm layout will normally have two components and they are called as *built up* area and *non-built up* area. Built up area consists of chicken houses like brooder, grower, layer houses and accessory facilities like feed mills, workers quarters, office, sales counter, godown etc. Non built up area consists of vacant land between two houses and other vacant land within the farm complex. Built up area in any farm is depends on flock size *i.e.* number of birds to be reared, floor space requirements, and area requirements for other facilities like feed mill, labour quarters, water tank, retail outlet if any etc., Non built up area mainly consists of vacant land, that is distance between two chicken houses, vacant land for future expansions etc.,

Construction Coefficient

Construction Coefficient is a useful tool to calculate the farm area requirement for a given farm size. It is defined as the Ratio of built-up area to the Total farm area and calculated as mentioned below:

Construction Coefficient = Built up area divide by Total farm area

The ideal Construction Coefficient for small farm size operations up to 30,000 layers is 0.40 - 0.50 and, for larger farm size operations it should be maximum of 0.30. Considering the construction coefficient, we can calculate the total farm area requirement for a given farm size.

Example: 1

A commercial layer chicken farm with 1, 00,000 square feet total area and 40,000 square feet built up area will have the Construction Coefficient of 0.40. (40,000 / 1, 00,000 = 0.40).

Example: 2

A layer farm with 30,000 square feet built up area will be required 1, 00,000 square feet total farm area for smooth functioning of the project and future expansions, with the construction co efficient of 0.30.

1. Built up area / Total farm area = Construction Coefficient
2. 30,000 / Total farm area (X) = 0.30
3. Total farm area (X) = Built up area (30,000) / Construction Coefficient (0.30)
4. Total farm area requirement : 30,000 / 0.30 = 1,00,000 square feet

Types of Poultry Houses

Broadly, commercial layer chicken houses can be classified in to two in two ways namely,

Deep litter house: It is a type of poultry house, in which birds are reared on litter material called as deep litter. It is providing a cushioning effect to the foot pad of birds and act as moisture absorbent material of the chicken droppings. Since, litter material is used in the floor of the house and it is also being added subsequently based on requirement, it is called as deep litter house.

Cage house: It is a type of poultry house, in which birds are reared in fabricated cages called as cage house. The house may be elevated or non-elevated cage house, depending on the floor height of the house. In elevated cage houses, cages will be set up in the first floor of the building, leaving the ground floor for collection of droppings.

Open sided house: Open sided house means, the sides of long axis of the houses will be open and closed only with chain link or mesh type material, allowing good cross ventilation. Since the sides are open, there will be limited control over micro environment and it is critically influenced by macro environment prevails outside the house. It is the most preferred housing system for Indian commercial chicken egg production system.

Environmentally controlled house: It is a house, in which sides will be closed with curtains like materials and inside house environment will be mechanically controlled by coolers, ventilators etc. Since the environment inside the house is controlled by artificial and mechanical means, it is called as environmentally controlled house. It is the housing system practiced in most of the developed countries. Considering its high investment and power requirement, it is not the preferred system in India.

Commercial Layer Houses

Commercial layer chicken can be grown in two systems namely deep litter houses or cage houses, which depends on the age of birds, interest and investment capability of the entrepreneur. The rearing of commercial layers is classified in to three phases namely, Brooders (0-8 weeks), Growers (9-20 weeks) and Layers (20- 72 weeks and above), consequently houses can also be classified in to three namely brooder, grower and layer houses.

Deep Litter Houses for Layers

Commercial layers can be reared in deep litter houses. The floor, feeding and watering space requirement for commercial layers is presented in Table 3.2.

Table 3.2 Space requirement of layers in deep litter rearing

Items	Brooders	Growers	Layers
Floor space (sq.ft)	0.5	1.0	2.0
Feeding space (cm)	2.5	5.0	10
Watering space (cm)	1.25	2.5	5.0

However, commercial layer chicken production mainly follows cage system of rearing, because of the many inbuilt benefits of this system for large scale operations. Considering the cost of investment per bird, operation ease in maintenance and level of automation, cage rearing is cost effective when compared to deep litter house for commercial operations.

Specifications for Deep Litter House

Orientation: The open sided deep litter layer houses should be constructed with their *Long axis* in *East - West direction* to avoid direct sunlight fall into the building. This type of direction will not only prevent entry of sunlight in the animal shed, but also prevent the splashing of rainwater during monsoon.

Length and width: The desirable range of width for open sided deep litter house is 20 -24 feet and width is always a limiting factor rather than length. Once the width is fixed, the length of the house is decided by the total floor space requirement and available dimension and topography of the land. If air movement is high in an area, width can be increased up to 28 feet. Otherwise, in hot and humid areas it is desirable to reduce the width to 24 feet to facilitate effective cross ventilation.

Floor: The floor of layer houses should be raised one foot (30 cm) above the outer ground level to prevent seepage of water into the house. The floor should be made of cement for easy and efficient cleaning, disinfection process and to prevent damage by rodents. It should be free from cracks and crevices. Improper flooring will leads to infection of micro-organisms, eggs of parasites, etc., which may cause outbreaks of diseases.

Walls: The height of long axis side walls of open sided poultry houses should not be more than one foot (30 cm) above the floor level and the rest of area up to eave should be covered with wire mesh (1.25 x 1.25 cm, 20-22 gauge thickness) or weld mesh (2.5 x 5.0 cm, 12-14 gauge thickness) or chain link (2.5 cm, 12 gauge thickness). The width of the side wall should be 6 - 8 cm and top of the sidewalls should be tapered and sloping inside called as *anti perch* slope to avoid perching of birds on the walls. The walls should be thoroughly cement plastered to avoid cracks.

Doors: The doors of open sided houses should be 3 - 3.5 feet width and 6 feet height and for every 40-50 feet length, one door should be provided with locking facilities from both sides. The door should be open outside of the house for working ease.

Roof: In open sided layer houses, the roof of may be thatched (coconut or palmyra leaves), Mangalore tiles, cement sheets (asbestos sheets), or new generation aluminum zinc alloy sheets (Galvalume). Thatched roofs are cheaper, ecofriendly and provides cool environment during hot season, but less durable is the disadvantage. Mangalore tile roofing is ideal for small and medium sized houses. Cement sheets (AC sheets) or Galvalume sheet roofing is preferred for larger houses, because of its durability and thermal comfort advantages. The height of the roof should preferably be 7- 8 feet at eave, and 11- 12 feet at the ridge. The roof pitch should be 3/12 or 4 / 12, which means three or four inch slope for 12 inches of *run* (run is half of the width of the house). Thatched roofs may have a lower height of 6 -7 feet at the eave and, ridge to eave slope should be steeper for easy drain of rain water. The extension of the roof at the eaves is called as *overhang* and should be at least 3 - 4 feet on either side to prevent direct sunlight and the splashing of rainwater into the buildings. The view of deep litter layer house is presented in Figure 3.1.

Figure 3.1 Deep litter chicken layer house

Feeding and watering space allowance: Feeders and drinkers are devices used to keep the feed and water for birds to take feed and water. Sufficient feeding and watering space per bird has to be provided, depending on their age, feed and water intake and climatic conditions. Commercially, feeders are available for chicks, growers and layers in various volume capacities and they may be used according to the age of the birds. Drinkers are also available in different sizes and volume capacities. They may be troughs or basins kept on the floor in the conventional manner, or hanging drinkers or fountain drinkers or nipple and cup drinker in automatic system.

The drinkers and feeders should be uniformly distributed over the floor area, so that the birds do not need to walk more than 150 cm. to reach a feeder or 200 cm. to a drinker. The height of the feeders and drinkers should be adjusted, so that the brim is at the same level of the back of the bird for easy pick up of feed and to avoid wastage and spillage. The feeders must be filled to only two-thirds of their capacity at any time to avoid feed wastage. The feeders and waters should be kept at the rate of one per 40-50 birds on any stage of growing irrespective of age.

Cage Rearing of Layers

Cage system of rearing is highly popular in commercial chicken egg production. It is the most predominant system of rearing in India and worldwide. In cage rearing, birds are reared in pre-fabricated cages with specific dimensions. Feeding, watering and egg collection arrangements are made in the cage itself for effective management. The common design followed in India is called as Californian cages.

The history of cage rearing dates back to 1930-1940 in United States of America. Initial attempts were made in the state California of USA through raising chicken on wire floor. Success of this system in improved livability, avoidance of pecking and fight for feed among birds and clean egg production leads the development of cage rearing worldwide. Since this system firstly followed in California of USA, these cage models are called as Californian cages.

Table 3.3 Recommended cage space allowance for chicken layers

Sl.No.	Age of the bird	Cage space (sq. inch)	Feeder space (inch)	Drinker Space (birds per nipple)
1.	Brooder (0-3 weeks)	15	2.0	4
2.	Brooder (4-8 weeks)	36	2.0	4
3.	Grower (9-20 weeks)	54	3.0	3
4.	Layer (21-72 weeks)	60	4.00	2

Advantages of Cage Rearing

The advantages of cage rearing system are many, which includes minimum floor space per bird, ease of automation for various farm activities, hygienic production environments, clean eggs, improved feed efficiency and livability, avoidance of pecking, and labour efficiency etc. Moreover, comparing deep litter rearing, cost of investment per bird in cage rearing is also less, because of its minimum cage and floor space requirements in cage houses. Considering that, all the large scale commercial farms are practicing cage rearing only.

However, the only disadvantage is raising concern over the issue of chicken welfare in cage rearing. In intensive poultry production, especially chicken layer farming, caging of birds is one of the most debated subject in terms poultry welfare. The reasons are obvious that, caging of birds prevents the birds to perform its normal behavioral characteristic like free movement, scratching, perching, and privacy of egg laying, feeding and watering space etc., making, caging a welfare issue. However, in deep litter system of rearing of layers, the floor space requirement is around two square feet and in cage system, the floor space allowance including cage space is around 0.6 sq.ft, which is 30 per cent of the floor space provided in deep litter rearing. The economy of floor space in cage system makes it popular for commercial rearing. Considering the volume of production, deep litter system of rearing may not be a reasonable and cost effective housing solution for industrial chicken layer farming. Though, there are different schools thoughts on welfare of cage rearing, Good Husbandry Practices in chicken egg production will be helpful in overcome the welfare issue by improved cage designs and cage space per bird.

Cage design and arrangement: Cages of various sizes are used to house birds during brooding, growing and laying. The cage will have three dimensions namely length, width or depth and height. Multiplication of length and width will provide the available floor space in the cage called as cage space. In layer cages, height of the cage will vary in back and front, making a slope for the egg to roll out of the cage. In Californian cage models, cages are arranged in two or three rows next to each other on either side. Conventionally, the bottom of the lower most cage is fitted at 75 cm height from the floor. Between cages, platforms are constructed for utility purpose. Feeders are fitted along the length of cage and cage length will be the available feeding space for that cage birds. Drinkers are in the form of nipples for easy watering of birds.

Figure 3.2 Open sided elevated commercial cage chicken layer house

External view

Internal view

Cages are fitted in two or three tiers on either side of the row under the Californian system. Two to three such rows of cages are arranged in a cage layer house. Based on the arrangement it is called as M or L type. Depending on the number of rows and the number of tiers in each row, the breadth of the caged layer house will vary from 40-50 ft. There is no stipulation for the length of such houses, which can be adjusted to the number of birds to be housed. No sidewalls are required for cage houses, as the mesh is stretched down to the floor level to facilitate better ventilation and to dry the moisture of the droppings. The elevated cage houses arrangement widens the gap between birds and their droppings, and facilitates quicker drying and easy removal of droppings. The graphical representation of open sided elevated cage house is presented in Figure 3.2.

Cage specifications: The cage specifications for various stages of birds like brooder, grower and layer will vary and the recommended cage dimensions are as follows,

Table 3.4 Recommended cage dimensions of commercial chicken layer

Sl.No	Item	Brooder	Grower	Layer
1.	Chicks per cage (Nos.)	12 Nos.	8 Nos.	5 Nos.
2.	Cage space per bird (sq.in)	36	54	60
3.	Cage front (inch)	24	24	20
4.	Cage depth (inch)	18	18	15
5.	Cage height (inch)	12	15	15/17*
6.	Feeding space (inch)	2	3	4
7.	Watering Space (No.)	4 Nipples	3 Nipples	2 Nipples
8.	Egg roll (inch)	Nil	Nil	7

*Height at back and front of the cage is 15/17 to provide slope for egg rolling out of cage for collection

Cage Houses for Chicken Layers

Two types of poultry houses are constructed for cage system of rearing in Indian conditions, namely floor level cage houses and elevated cage houses. The floor plan of elevated cage house and cage arrangement is presented in Figure 3.3.

Figure 3.3 Cage house floor plan and cage arrangement

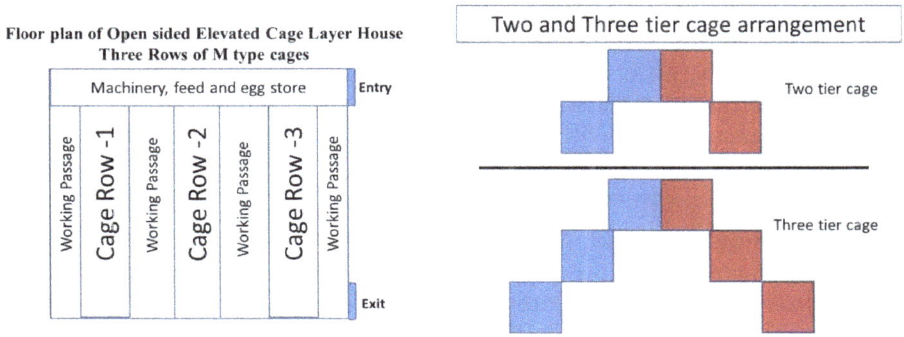

Cage house floor plan **Cage arrangement**

Floor level cage houses: Floor level cage house or non-elevated cage house is similar to deep litter house, except of deep litter floor rearing, birds are reared in cages. It is mainly followed in brooder houses of commercial layer chicken.

Elevated cage houses: In elevated cage houses, two floors are constructed and in ground floor, droppings will be collected and in the first floor birds will be reared in cages. Commercial chicken layer growers and layers are commonly reared in elevated cage houses. Cage houses may be of open sided cage house or environmentally controlled cage house. In India, considering the cost of investment, open sided cage house is the most preferred choice.

Elevated Open Sided Cage House Specifications

General specifications for open sided elevated cage houses for chicken grower or layer houses are,

1. Pillar elevation from the ground - 6 feet – considered as ground floor
2. Height up to ridge of the roof 14 feet
3. Pillar dimension of outer 12″ x 9 ″ and inner 9″ x 9″ with alternate pillars up to roof
4. Distance between pillars 15 feet
5. Width of the house - 42 ft to 50 ft based on number of rows of cages
6. Full or half monitor roof pattern
7. Over hang 5 feet
8. Roof material aluminum zinc alloy (Galvalume) sheets or cement sheets
9. Side cover chain link/ wire mesh 2 cm eye size
10. Working platform width 2-2.5 feet
11. Provision for feed and egg store at both the ends.
12. Provision of staircase to first floor with rodent control arrangement

Environmentally Controlled Cage Houses

In environmentally controlled cage layer houses, battery cages are being used for rearing layers to increase the stocking density per unit area of floor space. Battery cages are nothing but keeping cage boxes one above another (one row is called as deck) and providing automatic manure collection conveyor belts between cages. In contrast to open sided elevated cage layer houses, environmentally controlled cage houses are at ground level with suitable inner house height for accommodating the battery cages with required number of decks. Four or five decks battery cages are common model in commercial rearing.

In environmentally controlled poultry houses, the temperature and humidity remains consistent round the clock providing comfort environment to the birds. The reasonable comfort ambience for poultry rearing can be maintained in this house resulting in improved production efficiency, reduced disease incidences, with improved animal welfare in terms of comfort. The specifications for these type houses are usually recommended by the manufacturer of the house and cage system, as it needs careful research and development. The benefits of this system are increased stocking density, improved production efficiency, low incidence of diseases and mortality, automation and labour efficiency. However, initial investment per bird for these type houses are substantially on higher side. Considering the cost of production economics, profit advantage is only marginal over the open sided houses in Indian farming situations. Choice of this environmentally controlled house system for developing countries like India is questionable, especially, when there is scarce capital and electric power.

Open Sided Battery Cage Layer Houses

In India, recently attempts are being made with open sided battery cage houses by some entrepreneurs and getting positive feedback in the new system and its economic viability. It needs care and attention in cage and machinery design for automatic feeding, manure collection and ventilation. It offers increased stocking density, when compared to open sided Californian cage system of rearing.

Enriched Cages for Chicken Layers

Enriched cages are the large colony cages for 40-75 birds with more cage space per bird with the provision of nest, perch and litter materials for scratching, pecking and special devices to allow the birds to perform normal behavioral functions as like in the range system. European Union Council Directive 1999 prescribes standards for enriched cages. European Union imposed ban on (European Union Directive 1999/74/EC standards) conventional cages from the year 2012 in Europe and started implementing enriched and alternative production systems for commercial egg production. Model of enriched colony cages for layers is presented in Figure 3.4.

Figure 3.4 Model of enriched colony cages for layer chicken

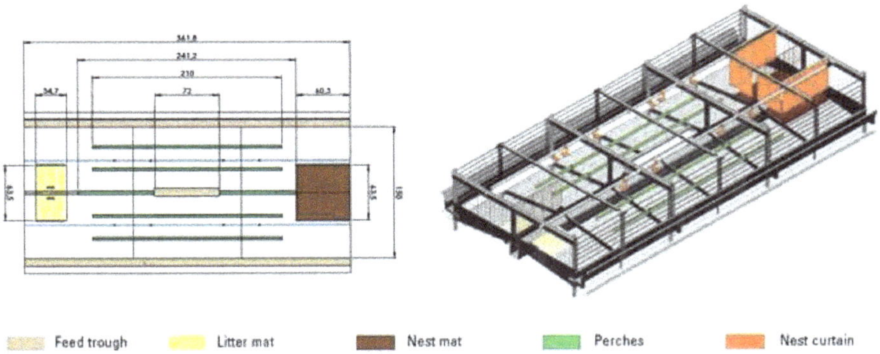

Source: https:// bigdutchman.com

Enriched cage specifications (European Union standards & specifications)

1. At least 750 square.cm (120 square inches) of cage area per hen, of which 600 square cm shall be usable.The height of the cage other than that above the usable area shall be at least 20 cm at every point and no cage shall have a total area that is less than 2000 square cm.
2. A nest
3. Litter for pecking and scratching
4. Perches allowing at least 15 cm per hen

5. Feed trough which may be used without restriction and length must be at least 12 cm multiplied by the number of hens in the cage
6. Each cage must have a drinking system for the group and if nipple drinkers are provided, at least two nipples or cups must be within the reach of each hen
7. To facilitate inspection, installation and depopulation of hens, there must be a minimum aisle width of 90 cm between tiers of ages and a space of at least 35 cm must be allowed between the floor of the building and the bottom tier of cages
8. Cages must be fitted with suitable claw shortening devices.
9. However, increasing cage space will have direct influence on production cost of eggs and enriched cage and alternative production systems will increase the production costs, which may not be preferred by the developing countries like India.

Activity

1. A commercial chicken layer farm's built up area requirements are as follows, and compute the total farm area requirement of farm with the construction co efficient of 0.25 and 0.30.

Brooder house	:	1 No. (8000 sq.ft)
Grower house	:	1 No. (15,000 sq.ft)
Layer house	:	3 Nos.(each measuring 18,000 sq.ft)
Mini feed mill	:	1500 sq.ft
Labour quarters	:	6 Nos.(each measuring 300 sq.ft)
Egg store	:	500 sq.ft
Office	:	300 sq.ft

2. Design a deep litter layer house for 5000 layers with suitable floor plan and cross sectional views.
3. Design a commercial chicken layer house for 30,000 chicken layers with suitable illustrations for cage design, floor plan and cross sectional views.

4

Layer Chicken Management

Learning Objectives

1. Describe the different phases and methods of rearing of layer chicken
2. Discuss the economic traits and the production performance of layers
3. Workout the production indices of commercial layer farm

Commercial Layer Chicken

Commercial layer chicken refers the hybrid chickens bred and reared exclusively for the production of table eggs for food purpose. They are the chicken, selected and bred for high egg production and low broodiness. The broodiness is the basic instinct of a chicken for natural incubation and reproduction. Broodiness and egg production are negatively correlated traits. Since, layer chicken are genetically selected for high egg production, it is almost free from broodiness. In their production process, at the hatchery itself, day old chicks are sexed by *Japanese vent sexing*. The female chicks only are sent to commercial layer chicken farming for further rearing and table egg (*infertile egg*) production for food purpose.

In this juncture, the student of veterinary and animal science should understand the difference between *fertile* and *infertile* eggs. Fertile eggs are the eggs, which possess developing embryo inside the egg. Fertile eggs are produced by the hens through natural mating or artificial insemination, when they are reared along with cocks. Infertile eggs produced by hens, when they area reared without males. Cock (male) is not necessary for a hen produce an infertile egg and when the hen attains physiological maturity, it will start lays the egg. Infertile eggs are used food purpose and cannot be used for hatching purpose since it devoid of developing embryo.

Phases of a Commercial Layer Rearing

Commercial layer chicken has three phases of rearing and they are called as brooding period (0-8 weeks), growing period (9-20 weeks) and laying period (21-72 weeks and above). The age at sexual maturity or *point of lay* in commercial layer is around 18-20 weeks. The egg production period is up to 72 weeks of age and above, thereby, egg production cycle is for one year and above (52 weeks and above). The system of rearing, feeding and management specifications is varying with the different phases of commercial layers.

Commercial Layer Farm Planning

Commercial chicken layer farming is a scientific rearing process, focusing on the maximization of production and productivity in terms of egg number per bird. Currently, the effective economic size of commercial layer farming is 50,000 layers per farm. Moreover, commercial layer farming requires a considerable investment per bird, which requires careful farm and financial planning.

Building plan: The design of buildings for rearing layer-type birds on deep litter is almost not in practice except of small country chicken egg production. The cage rearing of chicken layers of two types namely floor level and elevated cage houses. Floor level cage houses are mainly used in brooding of phase layer chicken. Elevated cage houses are mainly used for growers and layers. Considering the different phases in layer chicken, poultry houses may be classified as follows,

Brooder house: It is a house, in which chicken are reared from day old to eight weeks of age. They are floor level cage houses. The brooding of chicks is carried out in cages designed for that purpose.

Grower house: It is a house, in which, chicken are reared from 9-20 weeks period and they are elevated cage houses. Brooder cum grower houses models are also available in which chicken are reared from day old to 20 weeks in one house. This system is not common now in commercial operations.

Layer house: It is a house, in which layer chicken are reared during their egg laying phase (21-72 weeks and above) and they are elevated cage houses. In commercial layer chicken farms, egg production and supply will be of continuous process. To ensure continuous supply of eggs, proper plan is required in terms of constructing various number of different houses namely brooder, grower and layer houses and periodical inflow of day old chicks for rearing. The number of various houses required varies according to the batch interval of chicks. Based on this, the layer farm may be established as follows:

1 + 1 + 4 pattern: One brooder house + one grower house + four layer houses and batch interval for chicken placement will be 15 weeks.

1 + 1 + 5 pattern: One brooder house + one grower house + five layer houses and batch interval for chick placement will be 12 weeks.

1 + 1 + 6 pattern: One brooder house + one grower house + six layer houses and batch interval for chick placement will be 10 weeks.

Here, number of layer houses and its capacity indicates the egg production capacity of the farm. In a one lakh layer farm capacity with 1+1+4 house pattern, one layer house capacity will be 25,000 layers, accordingly four layer houses can accommodate one lakh layer birds at any given point of time. Individual batch size of 25,000 day old chicks at 15 weeks interval is required for continuous replacement of culled birds and to ensure continuous production of eggs.

Brooding Management

In natural incubation, brooding is a common activity of taking care of chicks by the mother hen in their early stages. In commercial poultry farming, brooding is an art and science of rearing a large number of day old chicks. It includes, providing enough warmth, feed and water in their young age usually two weeks. The brooding of young chicks is critical because of three reasons and they are:

1. A newly hatched out chick does not have well developed thermoregulatory mechanism to maintain uniform body temperature (*homeostasis*). This is because of its recent change over to *homoeothermic* from almost *poikilothermic* in egg incubation.
2. Feather is an excellent insulating material against heat loss but chick does not have fully developed feathers leads to heat loss.
3. The body temperature of chick is 41-42°C, which may cause heat loss in chicks. Therefore, it needs artificial warmth to get comfortable micro environment and to prevent death due to chilling of body.

Brooding period: In general, first two weeks of early life of chick is called as brooding period. Sometimes, it may extend up to third week, depending up on the climate of the area especially in winter. In extreme summer season, one week brooding will also serve the purpose.

Brooding Equipment

Brooding is of two types namely floor brooding and cage brooding. Floor brooding is mainly practiced in broiler rearing and cage brooding in commercial layer rearing. Brooding is an integrated setup, consists of artificial heat sources and chick guards or cages to keep the chicks warm. The feeders and drinkers are provided on the floor or in cages.

Floor Brooding

Brooder-heat source (gas brooder or electrically operated brooder or incandescent bulbs or infra-red bulbs or hot coal pot etc.). Hot coal pot is commonly used in floor brooding method in commercial broiler operations due to cost advantages. Chick guard – It is thin sheet, made of tin sheet or thick hard board or disposable paper pulp etc., to confine the chicks in a limited brooder area to provide adequate warmth to the chicks. Chick guard will always be arranged in a circular manner for the confinement of chicks. The height of chick guard will be 30-45 cm. Chick feeder and drinkers used for feeding and watering the young chicks during brooding period. The feeders may be circular or linear in shape. The waterers may be circular, nipple or cup type. Litter material is material used to spread on the floor to provide a cushioning effect to the foot pad of the chick and to absorb the moisture in the droppings. Old newspaper is spread on the litter material for the first week to prevent the eating of litter material by young chicks.

Preparation of brooding setup: Generally, preparation of brooding starts two days before the arrival of chicks. Complete cleaning, washing and disinfection of brooder house, three weeks before the arrival of chicks and keep the sheds empty for at least two weeks. Identify the brooding areas and it may be a part of shed (partial house brooding) or one corner of the shed (corner house brooding), which is depending on the need, convenience and brood size. Decide the size of brood and brood size should not exceed 400 chicks per brooding setup. Comfortable brooder floor space requirement is 200-225 sq.cm per chick (four chicks per sq. foot) for two weeks during brooding period. Based on the size of the brood (number of chicks in a brood) arrange the chick guard in a circular fashion and brooder heat source in the centre of the brooder area. Diameter of the circular brooder setup will decide the floor space available in the brooder setup and it can be calculated by using the formulae for area of a circle. Spread the fresh litter in the brooding area, one or two days before the arrival of chicks and spread of old and clean newspaper on the top of the litter. Keep the height of brooders and hovers one foot from floor on first day and slowly increase the height and it can be removed at the end of two weeks based on the room temperature. The cleaned and sanitized feeders and drinkers should be placed in brooders and they have to be arranged alternatively for easy availability of feed and water to chicks. The number of feeder and drinkers should be based on the number of chicks and at the rate of one feeder and drinker for 40-50 chicks.

Receiving of day old chicks: In general, day old chicks will be dispatched from the hatchery at evening and it will reach the farm at early morning to provide comfort during transport. As soon as the chicks are arrived in the farm, the baby chicks should be immediately transfer from chick box and placed under brooder on count basis and discard the chick box. Two hours before of arrival of chicks, heating source should be *on* in brooder and water should be filled in drinkers and do not fill up the feeders. The ideal first day brooding temperature is 35°C (95 ° F) and it should be reduced 0.5° C per day and 7th and 14th day it should be at 31°C and 28°C (88 and 82° F). The thumb rule for using incandescent bulbs is providing 1 watt per chick. It is always advisable to check the temperature at the brooder setup for maximum comfort. Once the baby chicks started drinking water, after an hour, feed shall be sprinkled in feeder trays and newspaper spread, to encourage the chicks to take feed. After that, regular feeders should be filled with feed. Here, it is important to mention that for the first two weeks, chicks should be given cooled boiled water with 10 gram of glucose with 1 g of electrolyte per litre of water as additives. Feed may also be offered in chick feed trays or chick box lids or egg filler flats to encourage the young chicks for better feeding.

Cage Brooding

Generally, commercial layer chicken brooding is carried out in floor level cage houses. The recommended cage space during this period is 15 sq.in for 0-3 weeks and 36 sq.in for 4-8 week of age. The number of chicks may vary from 12 to 50 per cage box. For first 3-4 days paper will be spread on the floor of cages to facilitate the grip for easy walk of the chicks. After that, paper shall be removed. To make the brooder house warm and conserve the heat inside the brooder house, the side curtains will be closed up to 80 per cent during this period. In commercial operations, cage fitted feeders and nipple drinkers are used for feeding and watering the birds. The recommended lighting program for brooding period is 22 hours from 0–3 days and 21 hours from four to seven days. Models of floor and cage brooding are presented in Figure 4.1.

Figure 4.1 Models of floor and cage brooding

Floor brooding

Cage brooding

Quality of brooding

Quality of brooding is assessed by the chick movement pattern within the brooder, feed intake of chicks, uniformity in growth and livability of chicks. Pattern of movements of chicks under brooder is the right indicator of the wellbeing of the chicks. The chicks should be active, move freely, scratch, take feed and water and make noise. If the chick distribution under the brooder is not uniform, correct the brooder temperature and ventilation accordingly. It is necessary to verify, whether the warmth given is sufficient for the chicks. A thermometer kept at the level of the birds will indicate the air temperature. However, a more practical way of assessing the adequacy of the warmth is by watching the behaviour and distribution of the chicks within the chick guard arrangement. If they crowd under or near the source of heat, the warmth is not sufficient. Then, additional heat may be added to the hover, or the height of the hover may be reduced. If chicks have moved to the periphery and are reluctant to come under the heat source, the temperature is too high. Then, the hover must be pushed up or a bulb shall be removed. If the chicks feel comfortable at the given temperature, they walk actively throughout the area and some will rest with their head on one side posture called "chick comfort". In practice, the heat source should be put on for 23½ hours in a day for the first 3 days, and switched off for only 30 minutes during night for adapting the chicks to dark. Later on, heat may be given during the night only, up to the end of the second or third week, depending on the season. It may be restricted to one week only during peak summer and extended to two to three weeks during winter or rainy seasons. In winter and rainy seasons, it is advisable to close the sides of the house with thick curtains to avoid cold drafts.

Assessment of the feed intake of chicks: Assessment of the feed consumption of chicks is highly essential to ensure the uniformity in weight gains in chicken. *Crop fill* test is an easy test assesses the feed consumption of young chicks. Collect 30 chicks randomly in a brood, and check the crop for its emptiness or fullness. Full and soft crop indicates the intake of feed and water and full and hard indicates the intake of feed and lack of water intake. The target for crop fill test is 100 per cent within 48 hours of brooding. The accepted brooding mortality in commercial layer rearing up to eight weeks period is five per cent. Crop fill test for chicks is presented in Figure 4.2.

Figure 4.2 Crop fill test for chicks

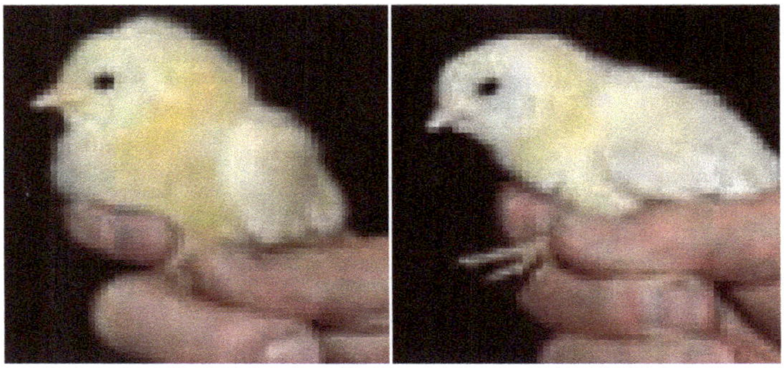

Chick with feed **Chick without feed**

Grower Management

The grower stage of layer chicken starts from 9 weeks of age and ends in 18 weeks of age. During this stage, the optimum body weight and growth of reproductive organs are critical to produce eggs in its egg laying phase. Growers will be reared in separate grower house. In case of cages, each grower is provided with 54 sq.in of cage space. During this period, grower diet should be provided with adequate protein and energy levels to support the growth of the birds. The important feature of grower feeding is feed restriction It is not advisable to provide *ad libitum* feeding at this age as the birds may tend to put on more fat and thus, their future egg laying ability may be affected. There are two types of restricted feeding adopted in growers.

Quantitative feed restriction: Quantity of feed is restricted to control the body weight and growth. This can be done on day-to-day basis programme. But this restriction depends on the matching of the flock average body weight with standard body weight provided by the breeder management guidelines.

Qualitative feed restriction: Quality of the feed is diluted in terms of protein, energy or diluted with high fibre. This can be done by manipulating protein, energy and fibre feed ingredients in the feed. Here the quantity of allotment to the bird is not restricted. In commercial operations, both qualitative and quantitative restriction is followed for better results.

Lighting: The recommended lighting program for growing period of 9-18 weeks is 12 hours. It means natural photo period of 12 hours is enough during this growing period. Do not increase the photo period during growing and it will lead to precocious sexual maturity.

Table 4.1 Mean feed intake and body weight of layer in growing period

Week	Feed intake / day (g)	Cumulative feed intake / week (g)	Body weight (g)
1	8	56	66
2	14	154	122
3	22	308	195
4	31	525	272
5	38	791	354
6	42	1085	441
7	45	1400	528
8	47	1729	615
9	49	2072	702
10	51	2429	784
11	53	2800	866
12	55	3185	948
13	57	3584	1025
14	59	3997	1102
15	62	4431	1169
16	66	4893	1225
17	72	5397	1266
18	79	5950	1292

Source: A Hendrix Genetics Company

ISA Management guide -www.isapoultry.com

Beak trimming: Beak trimming (often called as debeaking) is the removal of part of the top and bottom beak of a bird. It is a routine management procedure in commercial layer farms. The advantages of beak trimming in commercial farms are prevention of feather and vent pecking, cannibalism etc. and to minimize the wastage of feed. The beak trimming is carried out in the brooder and grower stage of layers. It is carried out in two stages and they are, first beak trimming at five to eight days of age and second beak trimming at 10-12 weeks of age.

Methods of beak trimming: There are two methods of beak trimming, which are called as hot blade beak trimming and Infra-Red beak trimming. In Indian farming practice, hot blade beak trimming is practiced in commercial farms. Hot blade beak trimming should be performed by trained staff. Two critical processes in hot blade beak trimming are Trimming and Cauterization of beak to arrest bleeding. Humane handling, right temperature of hot blade (Around 700°C – indicated by bright red colour of the blade), proper sizing in upper and lower beaks and cauterization of beaks after trimming are to be followed. The operator holds the chick in one hand with the thumb behind the head and the index finger under the beak. Slight pressure under the beak causes the bird to withdraw the tongue. Depending upon the age of chick, the upper and lower beak will be trimmed through the 4 mm opening. The cut will be about 2 mm in front of the nostrils. The cut surfaces of the beak can then be held on the hot blade for two to three seconds to cauterize and prevent bleeding. Beak trimming should be carried out under the supervision of veterinarian. Birds will find difficulty in taking feed immediately after beak trimming and it will become normal after 48 hours. Water should be made available to beak trimmed birds immediately. Supplementation of Vitamin K is beneficial after beak trimming.

Flock Uniformity

The purpose of feed restriction is to maintain uniformity in body weight of grower phase, there by attaining a desirable standard body weight at the end of the growing phase. Flock uniformity is measured by the comparing the body weight of individual birds with flock average. Here, 90 per cent flock uniformity is desirable. It means, 90 per cent of the birds body weight lies in the range flock average body weight plus or minus 10 per cent. It measured by using statistical tool called as Coefficient of Variation. Coefficient of variation is worked as Coefficient of Variation (CV) = Standard Deviation / Mean X 100. Here, good uniformity can be achieved through strategies such as receipt of chicks with uniform body weight, proper feeding, watering and floor space, proper energy in the diet and random weighing of pullets and checking with standards.

Layer Management

Laying phase of layer chicken starts from essentially from 18- 20 weeks of age and lasts up to 72 weeks and above. The extension of production beyond 72 weeks is based on the flock's egg production performance and market rate of egg. Growers should be transferred to layer cages at 17th week of age to adapt them for the new environment and lighting schedule. Grading and culling of poor layers can be practiced at the time of transfer from grower house to layer house optimize the cost economics in production.

Dynamics of Egg Production

The age at sexual maturity of commercial layer is 18-20 weeks; otherwise it is called as *point of lay*. During this time, birds will lay small sized eggs called as *pullet* eggs. Desirable body weight at the time of point of lay will minimize the number of pullet eggs. At the point of lay of the flock, pre layer ration shall be offered with incremental additional calcium level to condition the calcium reserve of the bird to support initial shell egg production. Then, there will be upward swing in egg production attaining a peak egg production of more than 90 per cent on hen day basis and remain there up to 45-50 weeks of age. Then, the production will start a down ward swing to 85-90 per cent production on hen day basis. The trait of egg production basically depends on number of factors namely age at sexual maturity, body weight at sexual maturity, intensity and persistency of lay, length of pause and age of depletion etc. Mean hen day egg number of 52 weeks production period will be 310 eggs. The mean daily feed intake per bird will be around 113-115 gram per day. The total feed intake during the growing period will be 42 kg. As the age advances egg weight will also increase, for which little bit of additional protein in terms of essential amino acids is required. Similarly as the age advances, egg weight and size will increase, consequently, reduction in egg shell thickness will be observed. Additional calcium will be required in the second phase of laying (above 45 weeks) to ensure egg shell quality. Because of these variations in production in laying, phase feeding is followed in commercial layers. The egg laying time is between 10 AM to 1.00 PM and eggs should be collected twice a day and summer thrice a day to maintain the quality of eggs. The collected eggs should be cleaned thoroughly and stored in egg filler flats with broad end up. The egg shell quality in terms of egg shell thickness will be of a problem during summer season and in aged birds. Poor egg shell quality will increase the incidence of broken or cracked eggs. The mean broken shell egg will be 0.5 to 1.0 per cent and the level of above this should be viewed carefully. Regular vaccination and medication if necessary has to be carried as per the advice of the poultry consultants. Dead birds should be collected on daily basis morning and evening and to be disposed scientifically after thorough post mortem examination

Peak egg production: The peak egg production is a period in which flock lays more than 90 per cent on hen day basis. It starts from 25-26th week of age and continues up to 48-50 weeks of age. Care should be taken in this period in terms of nutrition and management and any break in this period will affect the egg number very seriously.

Lighting management: Lighting is a critical aspect during laying phase, because it is a tool to optimize egg production and egg weight. The ideal photo period during laying period is 16 hours. However, from the 12 hours lighting period of growers, increase in lighting period should be smooth and incremental. From seventeenth week onwards increase the photo period to the tune of half an hour per week and end up in 24th week with a photo period of 16 hours. The recommended photo period 22-24 hours during brooding period gradually reduced to 12 hours at the end of eight weeks of age. Then the photo period should be kept constant for 12 hours up to 16 weeks of age. From 17th week onwards, the photo period of 12 hours per day is gradually increased to 16 hours per day with the weekly increment of half an hour per week up to 24th week. Then throughout the laying period, a constant photo period of 16 hours should be provided. The 16 hour photo period constitutes of 12 hours of natural and 4 hours of artificial photo periods. The four hour artificial photo period can be split in to two phases for morning and evening.

Culling: Culling can be applied, as and when necessary to remove the unproductive birds or at the end of laying period. The mean mortality during laying period is 10 per cent. The layers will be culled after 72 weeks and above and sent for table purpose as culled bird. Decision on culling age is not rigid, but flexible, which depend on factors like hen day egg production, feed cost, sale price of egg, availability and replacement pullet cost etc.

Table 4.2 Mean performance of commercial layer chicken (white egg shell)

Sl.No.	Traits	Performance
1.	Hatch weight	35-38 g
2.	8th week body weight	650 g
3.	Age at sexual maturity	18 weeks
4.	Body weight at sexual maturity	1.25-1.30 kg
5.	Age at five percent egg production	20th week
6.	Peak production period	26-50 weeks
7.	Hen Day egg number up to 72 weeks	310-320
8.	Hen Housed egg number up to 72 weeks	300-310
9.	40th week egg weight	56 - 58 g
10.	Feed intake up to 8th week	1.75 kg
11.	Feed intake up to 18th week	6.0 kg
12.	Feed intake during laying (19-72 weeks)	42 kg
13.	Mean daily feed intake during laying	113-115 g
14.	Feed efficiency per dozen eggs	1.6-1.7
15.	Feed efficiency per kg egg mass	2.3-2.4
16.	Livability during growing period	95 per cent
17.	Livability during laying period	90 per cent
18.	Age at culling	72 weeks & above
19.	Body weight at culling (72 weeks)	1.6 -1.7 kg
20.	Egg shell thickness	0.32-0.34 mm
21.	Haugh Unit Score egg at 40th week	86-88

Table 4.3 Economic traits of commercial layer chicken

Sl.No.	Economic Traits
	Pre laying period (Day old -20 weeks)
1.	Hatch weight – (Weight of Day old layer chick at the time of hatch after complete drying)
2.	Body weight at fourth, eighth, 12th, 16th, 18th ,20th weeks of age and at culling age (72 weeks and above)
3.	Body weight uniformity at eighth and 18th week - Expressed in Co efficient of Variation (CV)- High flock uniformity positively correlates with high egg production
4.	Feed consumption up to eighth and 20 weeks
5.	Percent culled outs during growing period (up to 20 weeks)
6.	Livability up to eighth week and 18th week
	Laying period (21-72 weeks)
7.	Age at Sexual Maturity (ASM) - Age of a bird in days when it lays its first egg Age at Sexual Maturity for a flock – Flock age at five per cent hen day egg production
8.	Body Weight at Sexual Maturity (BWASM)- Body weight of a bird when it lays its first egg

Table 4.3 contd...

Table 4.3 contd...

Sl.No.	Economic Traits
9.	Age at five per cent egg production – Age of a flock in days when it lays 5 per cent hen day egg production Age at 50 per cent egg production – Age of a flock in days when it lays 50 per cent hen day egg production Age at peak production - Age of a flock in days when it first lays more than 90 per cent hen day egg production Peak production persistency- Period of days, a flock maintaining more than 90 per cent hen day egg production
10.	Hen Day Egg Production(HDEP) – Number of eggs produced in a day for number of hens alive on that day (excludes mortality) Hen Housed Egg Production(HHEP)- Number of eggs produced in a day for number of hens housed initially in the house (includes mortality) HDEP, HHEP can also be worked out for the whole production period
11.	Weight of first egg Mean weight of egg at 40th week Mean weight of egg for the production period Egg mass during the production period (Number of eggs x Mean egg weight during the production period)
12.	Feed consumption during laying period - Total feed consumption from the point of lay to culling Average Daily Feed Intake (ADFI)- Mean daily feed intake during the laying period
13.	Feed efficiency per dozen eggs - Quantity of feed consumption to produce dozen (12 Nos.) of eggs Feed efficiency per kg egg mass - Quantity of feed consumption to produce per kg egg mass
14.	Egg shell thickness at 30,40,50,60,70th week of age – Thickness of Egg shell excluding shell membrane (expressed in mm) Broken eggs - Number of broken eggs and not fit for marketing (expressed in per cent) Egg abnormality (Meat spot, blood spot, misshapen eggs etc.,)
15.	Body weight of culled out for market (Weight of birds at the time final culling for meat market)
16.	Livability during laying period – Per cent livability from point of lay to 72 weeks and above or at culling.

Table 4.4 Production indices for commercial layer farming

Sl.No.	Production Indices
1.	Per cent Hen Day Egg Production for a single day (HDEP - Day) $\frac{Number\ of\ Eggs\ Pr\ oducedper\ day}{Number\ of\ Hens\ alvie\ on\ thatday} \times 100$
2.	Per cent Hen Housed Egg Production for a single day (HHEP-Day) $\frac{Number\ of\ Eggs\ Pr\ oducedper\ day}{Number\ of\ Hens\ Housed\ initialy} \times 100$

Table 4.4 contd...

Table 4.4 contd...

Sl.No.	Production Indices
	Per cent Hen Day Egg Production for a period
3.	$\frac{Number\ of\ Hens\ Housed\ initialy}{Number\ of\ Hens\ days\ during\ that\ period} \times 100$
	Per cent Hen Housed Egg Production for a period
4.	$\frac{Average\ Number\ of\ eggs\ produced\ during\ a\ period}{Number\ of\ Hens\ housed\ initially}$ ü 0
	Hen Day Egg Production for a period (HDEP- Period - Egg number)
5.	$\frac{Total\ Number\ of\ Eggs\ Produced\ in\ a\ period}{Average\ number\ of\ Hendays\ during\ a\ period}$
	Hen Housed Egg Production for a period (HHEP-Period - Egg number)
6.	$\frac{Total\ Number\ of\ eggs\ produced\ during\ a\ period}{Number\ of\ Hens\ housed\ initialy}$
	Feed Efficiency per Dozen eggs
7.	$\frac{Quantity\ offeed\ consumption\ (kg)}{Number\ of eggs\ produced} \times 12$
	Feed Efficiency per kg egg mass
8.	$\frac{Quantityoffeed\ consumption\ (kg)}{Number\ of\ eggs\ produced \times Mean\ egg\ weight}$

Activity

1. Collect the production data of a commercial farm and work out the following.
   - Age at sexual maturity
   - Body weight at sexual maturity
   - Hen day egg production in per cent and number
   - Hen housed egg production in per cent and number
   - Feed efficiency per dozen of eggs and kilogram of egg mass
   - Cost of feed per egg
   - Cost of production per egg

5

Commercial Layer Chicken Feeding

Learning Objectives

1. Discuss the role of various nutrients in chicken feeding
2. Discuss the commercial layer chicken feeding standards
3. List the common feed ingredients used in poultry feed
4. List out the critical steps in feed formulation of commercial layers

Role of Feed in Poultry Farming

Feed is the most critical input in commercial poultry farming system, and it is the single largest cost item in total cost of production of egg. In any industrial farming system, feed accounts for 65-70 per cent of the total cost of production. Quality of feed affects the production performance, and thereby cost of production. Escalation of feed prices due to cost of feed ingredients is alarming in recent past, creating lot of research focus on preparing cost effective or least cost feed. On the other hand, least cost feed may not be necessarily a quality feed always. The current trends on layer feeding is focusing the concept of cost of feed per egg should be the least, rather than least cost feed. Optimization of the feed formulae, there by maximizes the production efficiency is the base line in commercial layer feeding. In large and integrated poultry farm operations, feed is being prepared in their own facility in a way to reduce the feed cost with quality assurance. Considering this, a student of veterinary and animal science should understand the basics of layer chicken feeding as follows:

Basic function: Understanding the uniqueness in anatomical and physiological functions of digestive system of chicken.

Nutrient requirement of chicken: Nutrient requirements depends on chiefly, the type of bird, age of the bird like brooder, grower, layer, climatic profile of the area like tropical or temperate and bird's performance levels etc.

Feed ingredients: In industrial poultry farming, feed ingredients are more or less uniform throughout production areas. However, choices are inevitable, considering the availability, quality and cost factors, in which selection and quality control of feed ingredients is a critical process.

Feed formulation: Feed formulation is a scientific process, which needs understanding about the requirement of bird and nutrient profile of the feed ingredient. The poultry consultant as a feed formulator should know that, how to meet out the requirements by mixing various feed ingredients, and at the same time minimize the feed cost without compromising the feed quality and production performance.

Feed Milling: Feed milling is the actual feed preparation process in feed mills, here the challenge is to prepare a feed, may be mash or pellet or crumbles without affecting the quality of feed ingredients like nutrient quality reduction or wastage.

Feeding of poultry: It is the final step in feeding management, in which how to feed commercial chicken without wasting the finished feed is the challenge.

Digestive System of Chicken

The chicken's digestive system is simple, but efficient in function. The digestive system consists of alimentary canal, through which feed passes for digestion, metabolism, absorption and elimination. To support the digestion, accessory digestive organs like liver and pancreas produces bile and pancreatic juices which contain necessary enzymes for digestion. Digestive system of domestic fowl is presented in Figure 5.1.

The alimentary canal: The alimentary canal is tube like structure that, starts from the mouth and ends in cloaca. The nutrients from the feed, after digestion, are absorbed through the wall of the alimentary canal. The absorbed nutrients are further taken into the circulatory system for transport to the liver or other parts of the body. The waste remaining is eliminated through the cloaca.

Structure of mouth: Chicken doesn't have lips, teeth and cheeks. As prehensile organ, they have a hard beak, through which it is taking feed. Chicken do not have soft palate, making it difficult to suck water like other animals. Salivary glands in chicken are present in the whole of the mouth.

Oesophagus and Crop: The oesophagus in chicken is relatively longer and has a diverticulum or crop at the point of entry in to the thoracic cavity. Crop acts as a temporary storage organ or reservoir for feed.

Figure 5.1 Digestive system of domestic fowl

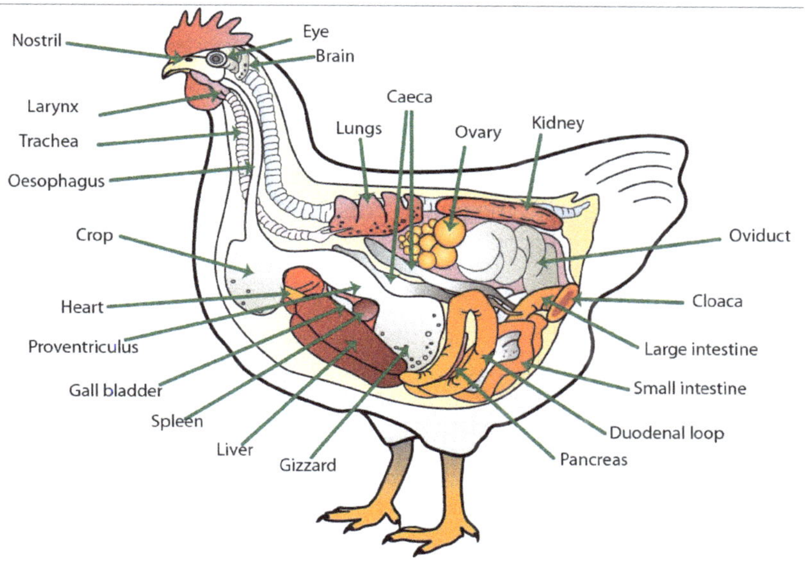

Source: http://www.poultryhub.org

Proventriculus: Proventriculus is an organ situated just before the gizzard, and at the end of oesophagus. It is small in size and holds little food material. The internal surface of the proventriculus is covered with glandular mucous membrane. It secretes gastric juice. Proventriculus is also known as *glandular stomach* or true stomach of the chicken.

Gizzard: The gizzard is muscular in nature called as *muscular stomach* of the chicken. It is the grinding chamber of the feed materials. The mucous membrane of the gizzard is coated with a thick layer, which is formed by the hardened secretions of gizzard called as *Koilin* layer.

Small intestine: The small intestine is the primary organ of digestion and absorption. It has been divided in to three segments viz. *duodenum, jejunum* and *ileum*. Duodenum originates from the distal end of the gizzard. It is difficult to differentiate the jejunum from the ileum in birds and these two segments are collectively called as lower small intestine. Most of the chemical enzymatic digestion and absorption and transfer of nutrients from intestine into circulatory system takes place in the small intestine.

Large intestine: Large intestine consists of caeca and colon. Caecum is a blind sac present at the junction of the small and large intestines. Colon is a short narrow tube extending from the *ileo - caeco-colic* junction to cloaca. The cloaca is the common passage for digestive, urinary and reproductive system. Microbial digestions of feed particularly of the cellulose containing materials are believed to be takes place in caeca. Absorption of water and soluble salts from digested feed occur in the colon region.

Pancreas: The three lobes of the pancreas located between the curve of the duodenum. Secretory ducts from the pancreas open into the distal end of the duodenal loop adjacent to the opening of the bile duct. Pancreas secretes pancreatic juices due to the stimulation of a hormone called *secretin*, produced from the small intestine.

Liver: Liver is the largest gland in then chicken body, and is the vital accessory digestive organ. The cystic duct is originating from the right lobe of the liver and gall bladder is the collection organ of liver secretions called as bile. Bile ducts open into the duodenum along with the pancreatic ducts. The primary role of liver in digestion is production of bile consists of bile salts and bile pigments.

Process of Digestion

The alimentary tract of birds is relatively short resulting a rapid pass of feed through the tract. The rate of passage of feed is faster in young birds (6 hours) than in the laying birds (8-9 hours). In broody or inactive fowls, the passage time may extend up to 12 hours. In chicken, the metabolic rate is high, which warrants a more or less a continuous supply of feed to the digestive tract and body system. When the birds are not eating, stored feed in the crop is used for the purpose. The ingested food is mixed with saliva and oesophageal mucus. Then feed enters the crop and moistened by the secretions. Amylase from the salivary glands and the enzymes present in the ingested feed cause hydrolysis of food. The proventriculus and gizzard are the first important sites of digestion on the food in chicken. Gizzard through its powerful grinding and mixing effect on the feed ensures sufficient exposure of peptide bonds to the action of the proteolysis. The pancreatic juices and bile secretions are added the ingested material at the distal end of the duodenum. The end products of the digestion of the ingested carbohydrates and proteins make their appearance in the circulatory portal within 15 minutes of the consumption. Pancreatic and intestinal lipases bring about lipid digestion in chicken. Bile salts helps in fat digestion. Most of the end products of digestion are absorbed from the small intestine. However, some absorption takes place from the caeca and colon also. The presence of villi and inner folding in the small intestine make it suitable for absorption by increasing the available surface area for the absorptive process. In the cloaca, faeces from the digestive tract are mixed with the urine and this mixture constitutes the droppings of the birds. An individual chicken dropping appears as a rounded light brown mass with a typical cap of white uric acid.

Nutrients for Chicken

Proteins: Proteins are the building blocks of the body. They are essential for proper growth and maintaining health. Most of the feed ingredients used in poultry feed contains protein, but their level vary with the type of feed ingredients. Cereal grains, the most popular poultry feed ingredient generally contain eight to twelve percent crude protein but cannot meet the bird's total protein requirement. It is therefore necessary to include other ingredients, which contribute higher levels of protein. Oil cakes are good sources of plant protein, while dried fish and meat meal supply animal protein. The amount of protein required by poultry varies considerably according to age and level of performance. There is a high demand for protein at the early stages of development. Protein is made up of several amino acids and it is necessary to know the composition of any protein in terms of its constituent amino acids profile. In commercial poultry production, protein requirements of the birds are not only expressed in terms of protein, but also at amino acid level. Based on that, amino acids can be classified in to three type, and they are as follows,

Non-essential amino acids: These are the amino acids, which can be synthesized from the available dietary protein sources and they need not be supplied in diet as special case and they are called as *dietary non-essential* amino acids.

Essential amino acids: The *dietary essential amino acids* cannot be synthesized by the body at adequate levels from the dietary sources. They have to be supplemented through diet based on the daily requirement. They are totally ten namely, lysine, methionine, threonine, tryptophan, isoleucine, leucine, histidine, valine, phenylalanine, arginine and glycine is for chicks only. Cysteine and tyrosine are considered *semi-essential amino acids,* because they can be synthesized from methionine and phenylalanine.

Limiting amino acids: These are the dietary essential amino acids, which are generally shortage in practical poultry diet. They are lysine, methionine and threonine and prevention of shortage of limiting amino acids in poultry diets is compulsory.

Carbohydrates and fats: Basically, birds require energy to maintain their body heat equilibrium to keep the body systems working continuously and to perform different types of activities. Birds derive their energy from two major groups of nutrients namely carbohydrates and fats. The requirements of carbohydrates and fats are expressed in terms of metabolisable energy (ME) requirement of a bird on daily basis. This is the reason, it is the most frequently used term to describe the energy value of poultry feed ingredients and compounded rations. The mean metabolisable energy values of common poultry feed ingredients are available for use in feed formulations. These values are calculated as gross energy minus foecal energy and gaseous energy losses, and expressed as kilocalorie per kilogram of feed ingredient.

Calorie Protein Ratio

Basically chicken eats for their energy need on daily basis. Consequently, the energy content of the feed determines the quantity of feed intake by birds. If the energy content of the feed is low, the quantity of feed consumption per day will be higher, and the other way round. Based on that, the intake of other nutrients in the feed like protein is also influenced indirectly by the energy content of the feed. There by, necessity arises to adjust the density of various nutrients of the feed, depending on the expected feed intake per day in given energy density of feed. Practically, most of the nutrients like vitamins and minerals provided in poultry feed will be excess or on par of the daily requirement except protein. Considering that, it is a regular process to adjust the protein content of the feed to the energy content of the same. This relationship is referred as *calorie protein ratio,* which is critical in real time feeding situations than the requirement of energy or protein as a single nutrient entity. It is obtained by dividing the energy value of the feed by protein value and expressed as ratio. As per BIS specifications, commercial layer feed in its laying phase -1 contains 2800 kcal per kg diet and 18 per cent protein. The calorie protein ratio is 155:1 (2800 divide by 18), which means, for 155 kcal intake of energy, the bird can take one gram of protein.

Vitamins: Vitamins are compounds, which play an important role in metabolism. Though, their requirement is very minimum, the presence or absence of vitamins makes production function high or low. Deficiency of certain vitamins is very critical, which can cause deficiency symptoms also. Good growth, egg production, prevention of leg weakness and thickness of eggshells, needs necessary vitamin levels in the diet. Apart from natural sources, they can also be added as supplements to meet the birds' requirements. Vitamins are classified into two namely, fat soluble and water soluble vitamins. The fat soluble vitamins are vitamin A, D, E and K and the water soluble vitamins are the B complex group and vitamin C. B complex group are thiamin (B_1), riboflavin (B_2), niacin, pantothenic acid, pyridoxine (B_6), biotin, choline, folic acid and cobalamine (B_{12}). Vitamins A, B_2 and D_3 are called as critical vitamins for poultry, as any deficiency will severely impairs the growth and egg production.

Minerals: Minerals are present in all body tissues and fluids, and perform many critical functions like the formation of bone and egg shell etc. Feed ingredients of animal origin contain more minerals than those of plant origin. The minerals required for poultry are classified as macro, micro and trace minerals, depending on their level of requirement. The macro minerals are calcium, phosphorus, potassium, sodium, and magnesium, the micro minerals are iron, manganese, zinc, copper and iodine, and the trace minerals are cobalt, fluorine, selenium, molybdenum, etc. As minerals found in common feedstuffs may not provide the bird's requirements, mineral supplements are added to the ration to overcome the possible development of deficiency symptoms

Water: Water makes up about 60 percent of the body composition of bird. Birds can survive without feed for two to three days, but suffer quickly without water with a threshold of 12 hours. A free supply of cool, fresh, clear, potable water should therefore be ensured in a day for efficient metabolic functions.

Nutrient Requirements

Basically, birds are fed in flocks; consequently, the nutrient requirements are not expressed on an individual basis, but expressed in unit weight of feed. Thereby, quantum of feed intake influences the nutrient intake and their balance with requirement. Nutrient requirements of birds are critically influenced by the age, sex, size of the bird, production efficiency, energy content, physical form of feed, nutrient availability, macro and micro environments. Therefore, recommending a common set of nutrients requirements for all birds is not practically possible. Allowance for variations in nutrient composition of feed ingredients, loss during storage, digestive efficiency of various nutrients should be offered to ensure a safety margin. Double the required amount is added for majority of the vitamins and minerals usually as a safety margin.

Poultry Feeding Standards

The poultry feeding standards prescribes the nutrient requirements of layer chicken during various stages of age, growth, production function to maximize or optimize the production efficiency. National Research Council (NRC) and Bureau of Indian Standards for Poultry feeds are common recommendations in poultry feeding standards. Recommendations of Bureau of Indian Standards (BIS-2007) shall be taken as a model (Table 5.1) for Indian climatic conditions. Slight modifications can be made based on the performance of commercial strains and the climate of the area in question.

Feed Ingredients

In commercial poultry industry, the feed is complete in nature, namely compounded and complete feed. It contains all the essential nutrients like energy, protein, fat, vitamins, macro and micro minerals and beneficial feed additives *etc.* to promote the production performance. Based on the feed ingredient's inclusion level in the feed, they can be called as major and minor feed ingredients. Commercial poultry feed is a balanced one, which can cater the nutrient requirements of poultry in all the aspects to optimize the economics of production and convenience of feeding. Based on the purpose of feeding, ingredients may be classified as energy sources, protein sources, mixed sources or filler materials, mineral, vitamin sources and feed additives. Common feed ingredients of layer chicken feed are presented in Figure 5.2.

Table 5.1 Bureau of Indian Standards (2007) - Layer chicken feed

Sl.No.	Nutrient	Type of feed			
		Chick (0-8 weeks)	Grower (9-20 weeks)	Layer Phase-I (21-45 weeks)	Layer Phase –II (46-72 weeks and above)
1.	Moisture (%), Max.	11	11	11	11
2.	Crude protein (%), Min.	20	16	18	16
3.	M.E. (Kcal/kg), Min.	2800	2500	2600	2400
4.	Ether Extract (%) Min.	2.0	2.0	2.0	2.0
5.	Crude fibre (%) Max.	7	9	9	10
6.	Calcium (%) Min.	1.0	1.0	3.0	3.5
7.	Avail. Phosphorus (%) Min.	0.45	0.40	0.40	0.40
8.	Lysine (%) Min.	1.0	0.7	0.7	0.65
9.	Methionine (%), Min.	0.40	0.35	0.35	0.30
10.	Methionine plus cystine (%) Min.	0.7	0.6	0.6	0.55
11.	Sodium chloride- Salt (%) Max.	0.5	0.5	0.5	0.5
12.	Acid Insoluble Ash (%) Max.	4.0	4.0	4.0	4.5
13.	Vitamin A (IU/kg) Min.	9000	8000	8000	8000

Table 5.1 Contd...

Table 5.1 Contd...

Sl.No.	Nutrient	Type of feed			
		Chick (0-8 weeks)	Grower (9-20 weeks)	Layer Phase-I (21-45 weeks)	Layer Phase –II (46-72 weeks and above)
14.	Vitamin D (IU/kg) Min.	1800	1600	1600	1600
15.	Choline (mg/kg) Min.	500	200	400	400
16.	Linoleic acid (%) Min.	1.0	1.0	1.0	1.0
17.	Aflatoxin B_1 (ppb) Max.	20	20	20	17

*All the prescribed values are on dry matter basis except moisture

Source: Bureau of Indian Standards (BIS) – IS 1374 -2007

Energy sources: These are the feed ingredients rich in energy value and the principal purpose of including these feed ingredients in poultry feed is for energy. Almost, all the cereals and fat and oil sources come under energy sources.

Maize (Corn): In Indian poultry feeds, maize is the most preferred and used cereal as energy source, because of its advantages like production, wider availability, digestibility, energy and cost factors. Quality of maize is decided by moisture content, bushel weight, fungus infestation and presence of foreign materials etc. The protein content of maize is around nine per cent with energy value of 3300 kcal per kg. Mostly, yellow maize is used in Indian poultry feeds, which is good source of *carotenoid* pigments called *xanthophylls,* imparts yellow colour to egg yolk and fat. It is also a good source of essential fatty acid *linoleic acid.*

Maize Gluten: Maize gluten is available in two forms namely maize gluten meal and maize gluten. Maize gluten meal is the by product in wet milling of maize and it contains 60 per cent protein and 3400 kcal energy. Maize gluten is similar to maize gluten meal and it contains around 20 per cent protein and both are good source of yellow pigments.

Jowar (Sorghum): Jowar or sorghum is a common rain fed cereal cultivated in monsoon seasons throughout India. There are two types of Jowar namely dark and white variety and both are used in poultry feeds based on availability and cost. Based on the cost factor, it can be used as replacement to one third of maize in poultry feeds. Tannin is the anti-nutritional factor present in sorghum especially in dark varieties; however, pelleting reduces the ill effects of tannin content.

Millets: The term millet refers to various minor cereals cultivated in India as rain fed crop. Commonly cultivated millets are pearl millet, finger millet, proso millet, foxtail millet, kodo millet, little millet and barnyard millet. The crude protein content ranges between 10 and 12 per cent with above 2500 kcal energy. Based on the availability and cost factors, they are used in poultry feeds to feed up to the level of eight to ten per cent.

Figure 5.2 Common feed ingredients of commercial layer chicken feed

Maize (Corn)	Jowar (Sorghum)	Bajra (Cumbu)	Broken rice

Soybean oil cake	Ground nut cake	Sunflower oil cake	Rice bran

Wheat bran	De oiled rice bran	Dried fish	Calcite powder
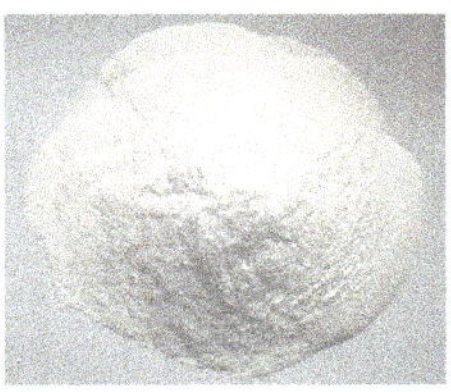			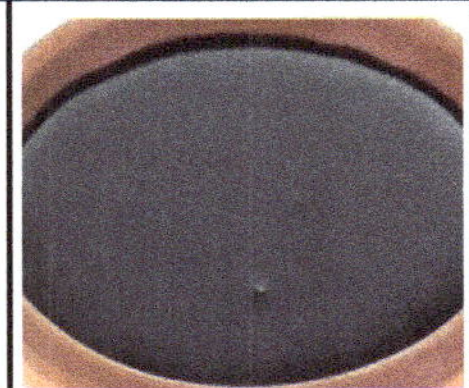
Di Calcium Phosphate	Shell grit	Meat meal	Molasses

Oats: Though oats is a good poultry feed ingredient worldwide, its use in Indian poultry feed is very limited, because of lack of availability and high fibre of around 12 per cent. Its use in high energy diets like broiler feed is not practical, because of presence of hulls and relatively low energy values.

Rice: India is one of the largest producers of rice and it is a staple food grain of people in major parts of India. Because of that, it's use Indian poultry feeds is very limited. However, broken rice is being extensively used in Indian poultry feeds, because of its availability and energy content. Moreover, use of good quality rice in poultry feeds should be avoided, because it makes poultry to compete with human beings for their food consumption.

Wheat: Production and consumption of wheat is worldwide in nature, and in India also, wheat is a staple food for people of Northern India. It is having lower energy value and higher protein content than maize. Wheat is used commercial poultry diets in partial replacement of maize, based on its availability and cost factors. Use of good quality wheat in poultry diets should not be encouraged because of the competition between human and poultry for food. The problem in using wheat is, its gelatinous nature making pasty vent conditions in practical feeding.

Barley: Though barley is produced and used in many parts of the world in poultry feed, its production and usage in Indian poultry industry is minimal. It is low in energy and high in fibre, when compared to maize, makes its use limited and uneconomical in Indian commercial poultry feeds.

Tapioca: Tapioca root (Cassava) is produced in many tropical countries including India. However, their use in poultry feed as energy source is very limited because of its anti-nutritional factors like *prussic acid* and special washing methods are necessary to remove these factors.

Molasses: molasses, a by-product of the sugar cane industries in India used as energy source which, contains three per cent protein. It is mostly used in pellet feeds as pellet binder and to reduce the dustiness of feed apart from flavor enhancer.

Fats and oils: Fats and oils of various types are used in poultry diets as energy source to improve the feed quality and palatability especially in high energy diets. It also offers the advantage of cost economics as energy source. Vegetable fats like coconut oil, palm oil, canola oil, rice bran oil, and animal fat like tallow or lard are used. Fats and oils are used in feed up to six per cent and it is also adding the benefit of cooling the birds, by releasing more metabolic water in their digestion and metabolism. Fats and oils are also potential source of linoleic fatty acid, which is essential in nature for poultry. The problem with adding fat in poultry feed is oxidative rancidity which can be prevented by adding antioxidants like *ethoxyquin* and *butylated hydroxy toluene* (BHT).

Protein sources: These are the feed ingredients rich in protein and used for the supply of protein to birds. They may be plant protein sources like all oil cakes or animal protein like fishmeal, meat meal etc.

Soy bean meal: Soy bean meal is one of the most preferred protein sources all around the world and in India for commercial poultry feed. The availability is also reasonably good in India, makes its use comfortable and economical. The protein content ranges from 40 - 50 per cent depending on the processing. Synthetic amino acids supplementation of lysine and methionine will very well balance the amino acid profile of the soy bean meal. There are three types of meal namely, expeller meal, solvent extracted meal, dehulled soy bean meal, which are having 43, 45 and 50 per cent crude protein respectively. Trypsin inhibitor is the anti-nutritional factor present in the soy bean meal, which is destroyed in heat processing.

Full fat soybean: Full fat soybeans are prepared by cooking, roasting or extruding methods and the purpose to increase the energy levels. It contains around 38 per cent protein and at the same time energy content is around 3200 kcal per kg. Heat treatment is necessary to destroy trypsin anti- nutritional factor, however in India, full fat soybean meal use in commercial poultry feed is limited.

Sunflower meal: Sunflower meal is an extraction byproduct in oil milling process of sunflower oil seed and commonly used in grower and layer chicken diets up to 15 per cent. The low protein of 25 per cent and high fibre of around 25 per cent limits its use in other poultry diets.

Linseed meal: Linseed meal (flax seed meal) is also good source of protein, but unpalatability of this product, limits its use in poultry rations seriously. On the other side, linseed oil meal is a rich source of omega-3 fatty acids and being used in rations to produce poly unsaturated fatty acids (PUFA) enriched special eggs.

Groundnut meal: Ground nut meal (peanut meal) is a good protein source for poultry rations, contains 40 - 45 per cent protein depending on processing. It can be used up to 15 per cent in poultry rations and easy contamination with mycotoxins is the problem in use of ground nut meal in stored poultry feed.

Rapeseed meal: Rapeseed meal (canola meal) is a good protein source with 38 per cent protein, but availability is less India. Inclusion level can be limited up to 10 per cent and higher inclusions will cause poor performance due the presence of *glucosinolate* and *erucic acid* like anti nutritional factors.

Safflower meal: Hulled and dehulled safflower meal is used in poultry diets and contains 27 and 32 per cent protein respectively. It can be used in adult poultry rations up to 15 per cent and of limited use in chick diets. It is high fibre of 14 per cent limits its use in commercial poultry diets.

Coconut meal: Coconut meal (copra meal) is the byproduct of coconut oil extraction process and the mean protein content is 20 per cent with high fibre content of 14 per cent. The inclusion level in poultry diet may be up to eight per cent.

Cottonseed meal: Cottonseed meal and dehulled cotton seed meal are good quality protein sources with about 40 and 45 per cent protein. However, their use in poultry diets is limited because of *gossypol* factor, which is toxic and causes changes in yolk colour and growth reduction.

Fish meal: Fish meal is dried and ground fish used in commercial poultry feed. Sun dried fish is commonly used product in Indian poultry feeds. It is a good source of protein to poultry, because of its balanced amino acid profile. The protein content of fish meal varies from 40 - 50 per cent based on its quality. The energy of the fish meal depends on its fat content. The problem with fish meal is its salt content and it ranges from 4-7 per cent and it depends on processing methods. Considering that, sample should be tested for every batch of product before including in the rations. Good quality fish meal with three to four per cent salt can be added up to eight to ten percent in commercial poultry diets.

Blood meal: Blood meal otherwise called as dried blood contains around 75 per cent protein, rich in lysine but totally of poor protein quality. It is used in poultry diets with very minimum inclusion levels.

Meat by products meal: Meat meal and meat and bone meal are being used Indian poultry feeds, of which major is the meat and bone meal. Meat meal as protein source (50 per cent) used in poultry diets at the inclusion level of five to six per cent. Meat and bone meal is a protein (40 per cent) and mineral source used in poultry diets up to eight per cent. The major problem in using meat by products meal in poultry feeds is risk of Salmonella contamination.

Poultry by products meal: It is a product of rendered poultry offal in dry form which includes head, intestines, and visceral organs, excluding feathers. It contains 55 per cent protein and 12 per cent fat. It is considered as an excellent source of protein for both broiler and layer diets. However its use in commercial poultry feeds is limited because of lack of availability.

Hydrolyzed poultry feather meal: Feather of chicken broiler, layer turkey are steam cooked at high temperatures, dried and ground to prepare hydrolyzed feather meal, contains almost 70 per cent protein. However its use in poultry rations is limited and up to eight to ten per cent in replacing vegetable protein sources due to digestibility problems and poor quality amino acid profile.

Mixed sources: These are the feed ingredients used in poultry feeds as energy and protein sources and sometimes called as filler materials to reduce the cost of feed.

Rice bran: Rice bran is the fine outer covering of rice germ and it is a byproduct in rice milling process. It contains 12-14 per cent protein and around 2800 kcal energy per kg, make it a potential poultry feed ingredient as energy, protein and filler sources. However, because of its relatively high fat content of around 14 per cent, it poses rancidity problems in storage. Nowadays, oil is extracted from rice bran as rice bran oil and used by people for edible purposes. The byproduct deoiled rice bran is highly used in poultry feeds as protein source and filler material, which does not have rancidity problems in feed storage.

Wheat bran: Wheat bran is similar to rice bran, outer layer of the wheat kernel and a byproduct of wheat milling process. It contains 15 per cent protein and low energy value of 1300 kcal per kg when compared to rice bran. It is also being used in poultry feed extensively because of its wide availability and cost factors. One of the other by product of wheat milling used in poultry diet is wheat middlings, and they are mix of bran, broken germ, flour etc. contains 16 per cent protein and 1800 kcal energy per kg.

Mineral sources: These are feed ingredients used in poultry feed to supply mainly macro minerals to birds namely calcium, phosphorous, sodium and chloride.

Dicalcium phosphate: The base source of dicalcium phosphate is rock phosphate or bone and through chemical processing dicalcium phosphate is prepared. The adulterant in dicalcium phosphate derived from rock phosphate is fluorine, which has to be removed before use. Dicalcium phosphate contains 22 and 18 per cent calcium and phosphorous.

Defluorinated rock phosphate: Defluorinated rock phosphate rock is a ground product of raw rock, which is defluorinated by subjecting to high temperatures to remove the high fluorine content. It contains about 32 per cent calcium and 8 per cent phosphorus.

Bone meal: It is a steamed and ground product of animal bones, which contains 30 per cent calcium and 12 per cent phosphorous.

Limestone: Lime stone is an exclusive calcium source contains 38 per cent calcium and used as calcium source for commercial layer chicken feeds.

Oyster shell: Oyster shell is a source of calcium because of its calcium availability and particle size and contains 38 per cent calcium.

Salt (Nacl): Salt (Sodium chloride) is a source of sodium and chlorine required in minor quantities in poultry feed and to improve the palatability. Deficit and excess will cause harmful effect to the birds.

Alfalfa leaf meal: The are many grasses and legumes are dried and fed to poultry to act as source of beta carotene, xanthophylls like pigments and un identified growth factors. The commercially important product is alfalfa leaf meal. It is the sun dried and ground alfalfa leaf with moisture of less than 12 per cent. However, use of this product in commercial poultry feed is decreasing because of availability of synthetic products at competitive rates.

Feed Additives

A feed additive is a substance or mixture of substances added to the feed especially in very small quantities apart from regular feed ingredients. A feed additive may have nutrient and non-nutrient functions. Functions of feed additives are to improve the nutritive and feeding value of the feed and enhance the taste, flavor, palatability and prevent the spoilage of feed.

Table 5.2 Inclusion levels and nutrient composition of feed ingredients

Name of the ingredient	Inclusion levels	C. P. (%)	M.E. (Kcal/ kg)	C.F. (%)	Ca (%)	NPP (%)	Lysine (%)	Methionine (%)
Maize	60	8.9	3300	2.3	0.03	0.1	0.22	0.18
Jowar	30	10.3	3256	2.4	0.05	0.011	0.21	0.16
Bajra	30	11.6	2850	4	0.05	0.12	0.24	0.16
Wheat	30	11.0	3120	3	0.05	0.1	0.33	0.18
Broken rice	40	8.5	2900	3	0.08	0.1	0.21	0.14
Rice bran	30	12.9	2400	12	0.05	0.22	0.39	0.13
Deoiled rice bran	40	13.5	1900	14	0.07	1.01	0.62	0.28
Vegetable oil	06	-	7800	-	-	-	-	-
Soybean meal (SE)	40	42.0	2420	6.5	0.29	0.26	2.59	0.52
Groundnut meal (SE)	30	43.0	2100	12.2	0.17	0.18	1.61	0.44
Sunflower meal (SE)	30	27.0	1900	24.9	0.37	0.91	1.14	0.65
Fish meal – 45%	10	45.0	2460	1	5.39	2.99	2.27	0.87
Meat meal – 50%	04	50.0	2200	2	6.27	3.1	2.61	0.69
Di-calcium phosphate	04	-	-	-	22	18	-	-
Calcite	04	-	-	-	36	-	-	-
Shell grit / lime stone	04	-	-	-	38	-	-	-
SE- Solvent Extracted								

Nutrient feed additives: The nutrient feed additives are the substances that contain essential nutrients necessary for the normal growth and production of birds. The nutrient feed additives can be further classified into three namely, amino acid, vitamin and mineral supplements that contribute essential amino acids, fat and water soluble vitamins, micro and trace minerals etc.

Non-nutrient feed additives: Non nutrient feed additives are the substances which do not have any nutrient role but it will improve overall quality of feed by various functions. The commonly used non-nutrient feed additives are,

Feed Enzymes: Products enhance the digestion and metabolism of the various feed ingredients. They are the enzymes like *protease, lipase, cellulase, amylase, phytase* and *pectinase* and mixture of these enzymes like Non Starch Polysaccharide (NSP) enzymes.

Anti-oxidants: Substances that prevent oxidative rancidity of fats and oils. They are mainly Vitamin E, *Ethoxyquin, butylated hydroxy toluene* (BHT) and *butylated hydroxyl anizole* (BHA).

Adsorbents: Substances that absorb or bind toxins and prevent their absorption from the intestine like zeolites, activated charcoal.

Pellet binders: They are used in pelleting process of feed for the preparation of pellet or crumble feed for chicken. The common pellet binders are *bentonite,* sodium alginate, *carboxymethyl cellulose, gelatine, lignosulphonate and guargum.*

Probiotics: The probiotics are the culture of beneficial and nontoxic microorganisms, which produces beneficial effects to gut and animal health. These are made from bacteria, yeast and fungi. Commonly used probiotic preparations are bacterial strains of *Lactobacillus, Leuconostoc, Bifidobacterium, Pediococcus* and *Streptococcus*, fungal strains of *Aspergillus* and Yeast strains of *Saccharomyces.*

Prebiotics: It is a non-digestible feed ingredient that promotes the growth of beneficial microorganisms in the intestines. The prebiotic is a non-digestible carbohydrate commonly oligosaccharide (Fructose or Mannan oligosaccharides - FOS or MOS) which has two functions like supplying nutrients to beneficial microbes and attaching pathogenic bacteria and prevent them to attack intestinal mucosa.

Flavouring agents: Substances that improve feed flavour thereby feed intake, growth rate and production. Examples: Vegetable oils, fish oils, etc.

Pigments: Substances that impart an attractive colour to the products like egg yolk and chicken skin. The common products are *canthaxanthin, leutin, zeaxanthin,* etc.

Deodorising agents: Agents which reduce the ammonia production in the litter like *yucca* plant extracts.

Coccidiostats: They are the drugs prevent outbreaks of sub clinical coccidiosis. The common products are monensin, amprolium, salinomycin, nicarbazine, and maduramycin.

Antibiotic supplements: They are used to control subclinical bacterial infections and thereby to boost performance. They are mainly tetracyclines, lincomycin, erythromycin, colistin, doxycycline, bacitracin, flavomycin, virginiamycin. However, use of antibiotics is very limited in current situations, instead use of pro and pre biotics are extensive in poultry feed, due to the growing concerns over the issues of antibiotic resistance and food safety.

Feed Formulation

Generally, commercial poultry feed formulation is done by nutritionist, specialized in poultry science by using principle of least cost feed formulation with the use of computer. In earlier days Pearson square methods, equation methods are used to fix up the energy and protein levels. Now, computers are extensively used in feed formulation process. The feed formulator should have thorough understanding about the nutrient requirement of various type of chicken like layers, nutrient composition, inclusion level and cost factors of various feed ingredients to prepare a balanced, least cost, and quality feed. Basically, the poultry feed consists of 50-60 per cent grains as energy source, 25-35 per cent oil meals, fish and meat meal as protein source, 25 per cent filler materials and three to four per cent mineral sources. Additives will be added as per the requirement. Addition of molasses will improve the palatability, flavor of the feed, reduce the dustiness and improves the pellet quality. Molasses and oil incorporation in the feed offers beneficial effect during summer. Addition of oil or fat will the increase the energy level of feed will provide cooling effect due to higher metabolic water yield, increased energy concentration will reduce the feed intake and improves the feed efficiency. Marginal deficit of lysine, methionine and threonine can meet out by addition of synthetic preparations. In practical feed formulations, if the levels of protein, energy, lysine and methionine are achieved, practically, there is no need to check other nutrients and mostly deficit will not occur in other nutrients, even if is so it can be met out by adding synthetic feed additives.

Poultry Feed Milling

Poultry feed milling is mostly an engineering process, which needs full fledge feed mills. It consists of batch weighing scales, grinders, mixers, packaging machines and quality control lab for testing finished feed. Commercial layer chicken farms operate their feed mill with experienced feed mill experts and poultry consultants for better results. Feed cost shall be reduced considerably in own feed preparation; however the farmers should be financially capable of bearing the feed ingredients inventory. When feed ingredients are available locally, own feed preparation will be profitable. Large commercial farms are capable of investing more on ingredient procurement, feed preparation and milling operations because of their volume, which offers substantial reduction in cost of feed. They can make use of high tech milling operations, resulting in quality assurance. Schematic diagram of feed milling process is presented in Figure 5.3.

Quality Control in Poultry Feed

In poultry feeding, the performance of the birds depends upon the quality of the feed and the feed should be fresh, nutritionally balanced and free from other contaminants like fungus growth etc. Selection of feed ingredients is critical for preparing quality feed and quality control starts from procurement of feed ingredients itself. Good quality, fresh feed ingredients and feed supplements must be used in the feed preparation. The feed and feed ingredients must be well stored for a shorter period, with less than 10 per cent moisture and free from fungus growth. Use of right feed formulae and correct weighing of raw materials as per the formula is essential for quality feed preparation. Proper grinding, mixing and adding of feed additives will ensure the quality of finished feed.

Figure 5.3 Schematic diagram of feed milling process

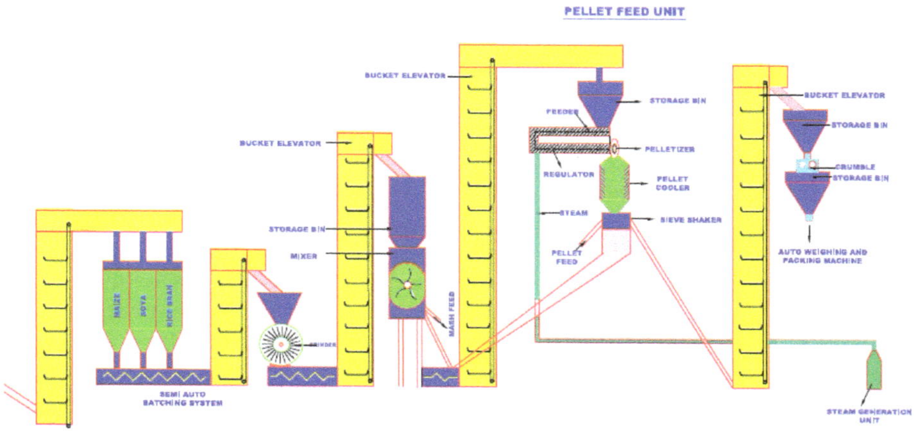

Source: Dept. of Animal Nutrition, VC&RI, Namakkal, TANUVAS, Tamil Nadu

Quality testing of feed ingredients: Basically, all the feed ingredients should be tested physically by organoleptic tests and then chemically in a quality control laboratory for their quality standards. Materials with objectionable standards should be rejected.

Physical examination: In physical examination, the feedstuffs are subjected to visual examination under bright sun light for colour, freshness, dryness, caking, mould growth, insect attack, adulteration etc. It can also be smelled for off odour, rancidity, and other abnormal odours. Thus, physical examination will be a good and reliable indicator of the feed ingredients quality at field level to take immediate buying decisions without any laboratory help.

Laboratory examination: For large scale feed mill operations, laboratory testing of feed ingredient quality is mandatory. Physical and laboratory tests are not mutually exclusive and they are complementary in nature in improving the overall quality of feed. In addition to the physical tests, recommended laboratory tests for feed ingredients are as follows:

Table 5.3 Recommended quality control tests for various feed ingredients

Sl.No.	Type of feed ingredient	Type of quality control test
1.	All the grains	Estimation of moisture and Aflatoxin
2.	All the oil cakes	Estimation of crude protein, Aflatoxin
3.	Rice bran and polish	Estimation of sand and silica and fat
4.	De oiled rice bran	Estimation of protein, sand and silica
5.	Fish and meat meal	Estimation of protein, sand and silica
6.	Mineral mixture	Estimation of calcium, phosphorous and flourine
7.	Di calcium phosphate	Estimation of phosphorous
8.	Soy bean meal	Protein, Urease and Trypsin inhibitor test
9.	Rock phosphate	Phosphorus, fluorine

Forms of Poultry Feed

Commercial poultry feed is prepared in three forms namely, mash, pellets and crumbles. In India, almost all the commercial broiler feeds even for various phases are prepared in the form of crumbles and pellets. In layer side, only chick feed is prepared in crumble form and other grower and layer feeds are mostly in the mash form. Forms of poultry feed is presented in Figure 5.4.

Figure 5.4 Forms of poultry feed

Mash **Crumble** **Pellets**

Mash type: In mash form of feed, different feed ingredients are ground in coarse particle size and mixed to form a complete ration. However, fine coarse particle is prepared when it is meant for young chicks. The disadvantage is some amount of selective feeding is possible and dustiness of the feed in use and transportation. Addition of molasses and fat will become a problem in mash feed.

Pellet type: Pellet is a compressed form of mash feed. The mash is compressed using an equipment called pelleting machine to form pellets of various sizes. The pelleting machine die with holes of a specific diameter, through which, the steam heated feed is forced under pressure to form pellets. In pellets, selective feeding is not possible and highly suitable for young chicks and high energy diets like broiler diets. Fat may be added in larger quantities to pelleted feed by spraying it, after the pellet is made. The process of pelleting improves palatability and mostly increases nutrient availability of diet by denaturing protein, gelatinizing carbohydrates and destroying trypsin growth inhibitors. The advantages of pellets are generally high and make it more suitable for commercial poultry feeding programs.

Crumble type: When coarsely ground pellets are pressed and run through special cracking machine, midway product between mash and pellets is formed known as crumbles. Because of the smaller size, it may be fed to young chicks.

Phase Feeding

The basic principle of phase feeding is, layer birds will have different nutrient requirements during various stages of age, growth and production. Formulating a single feed for all the stages is not logical and economical in terms of efficiency of feeding. Based on that, different feeding standards are prescribed for different production functions and its efficiency of production. In commercial layer production, phase feeding is practiced in various stages of layer chicken.

Table 5.4 Phase feeding in commercial layer chicken rearing

	Chick feed (0-8 weeks)
	Grower feed (9-18 weeks)
Layers	Pre layer feed (19-20 weeks)
	Layer feed - Phase -1(21-45 weeks)
	Layer feed - Phase -2 (46 weeks and above)

The main objective of phase feeding is to minimize the wastage of nutrients by over feeding than the bird's actual needs under different stages of growth and production to obtain cost effective production function.

Pre lay diets: Pre lay diets are designed to grow the bird to store adequate calcium in medullary bone reserves which is necessary for the birds for calcification of the first egg. Generally, these diets fed from two weeks prior to egg production and up to five per cent hen day egg production. Pre lay diets generally contain 2-2.5 per cent calcium. Lack of good pre lay diets may result in lameness, cage layer fatigue and drop in egg production.

Restricted feeding: The concept of restricted feeding is imposing feeding restriction to the birds to control their growth and production function. It is just opposite to *ad libitum* feeding in broilers. It is highly practiced in broiler breeders to restrict their growth during their growing phase to have a control over its body weight and to achieve a desirable body weight at sexual maturity. The desirable body weight at sexual maturity is a critical factor, which influences the hatching egg production substantially. Since, the broiler parents are selected for high growth rate, over intake of feed and overweight will be a problem in its growth stage, which may lead to early sexual maturity and low egg production. However, in layer feeding feed restriction is mildly practiced during grower stage to avoid excess body weight. Commercial layer feeding experiences indicates that, achieving desirable daily feed intake and body weight at sexual maturity itself is a problem rather than excess feed intake in layers. Restricted feeding is of two types' namely qualitative and quantitative feed restriction. In qualitative feed restriction, dilution of feed quality is employed by reducing the protein, energy or increasing the fibre levels there by, imposing a restriction on nutrient intake. In quantitative feed restriction, feed allowance per day per bird is restricted to have a control over actual nutrient intake. In commercial layer feeding, both qualitative and quantitative feed restrictions are practiced at mild level with daily feed restriction to achieve good flock uniformity in body weights.

Summer Management of Layer Chicken

In Indian climatic conditions, it is common to have high ambient temperature and in some areas often accompanied with high humidity. In these situations, acute heat stress will substantially affect the overall productivity of a layer flock. In environmental temperatures above 35°C (95°F) heat stress induced mortality and marked production losses are common. However, at lesser temperatures; heat stress is often viewed as a predisposing cause for chronic loss of production layers. Clear understanding of physiology of heat stress and managemental strategies to overcome the same is highly essential for tropical chicken egg production systems.

Physiology of thermo regulation in chicken: Chicken is a homoeothermic animal, and they try to maintain a relatively constant deep body temperature of 41-42°C (105 to 107) irrespective of environmental temperature. The thermo neutral zone or *Comfort zone* is a particular range of environmental temperature, in which, birds do not change their behaviour or show signs of discomfort and use minimum amount of metabolic energy to maintain normal range of body temperature.

In thermo neutral zone, body temperature is maintained by the thermal equation as heat production is equal to heat loss. Increasing environmental temperature causes hyperthermia. Action on reducing hyperthermia can be activated by increasing heat dissipation and or decreasing heat production. Chicken, unlike most other animals, do not have sweat glands to aid heat loss. The chicken removes excess body heat in four ways through the signals of thermoregulatory centre located at hypothalamus. Body heat can be lost by *radiation* from the skin surface through the air to another object. Heat can be directly transferred by *conduction* to cooler objects with, which the bird is in contact, such as litter. Body heat is also lost to the surrounding air by *convection*. When the environmental temperatures are between 30°C and 35°C (86°F and 95°F), radiation, conduction, and convection heat losses (*sensible heat losses*) are usually adequate to maintain the bird's body temperature. In these temperature zones, the bird dilates the blood vessels of the skin, wattles, and comb to bring the internal body heat to the skin surface to facilitate conductive, convective, and radiative heat losses. Birds at floor will search cool places in the house, spreads the wings and squats to increase the conductive and convective heat loss by increasing the body surface area.

When environmental temperature approaches the body temperature of the bird, 41°C (105°F), the efficiency of these heat loss mechanisms diminish. At this point, evaporative heat loss (*insensible heat loss*) that is the evaporation of water from the respiratory tract and mouth becomes the major heat loss mechanism of the bird. Birds will begin to pant (open mouth breathing), or hyperventilate to increase evaporative cooling. When panting fails to prevent the rise in body temperature, the bird becomes listless, comatose, followed by death due to heat stroke. Birds raised from young age at high environmental temperatures will acclimatize to higher environmental temperature variations and extremes and can maintain good productivity even in high ambient temperatures to a certain extent.

Effects of heat stress: One of the primary effects of high environmental temperature on a flock is reduced feed intake. Reduction in appetite is the birds' effort to reduce energy intake in response to the increase in environmental temperature by neuro signals from hypothalamus and thyroid, thereby reducing the energy obtained from feed. Birds may use body fat as a source of energy that produces less heat increment than metabolism of proteins or carbohydrates of the feed. The reduced feed intake and subsequent deficit of other critical nutrients quickly affect the productivity of the flock. Drop in egg production occurs in layers with poor egg shell quality. The nasal passages function is to filter dust and bacteria from air entering the respiratory tract, and this system is by-passed, when birds resort to open-mouth breathing (panting) leads to easy entry of infectious agents.

Nutritional management: Heat-stressed flocks usually experience a loss in appetite. This reduced feed intake can be compensated by formulating a nutrient rich feed. The actual energy requirement of the bird is reduced at high environmental temperatures, because the birds do not require energy to maintain its body temperature, but all the other dietary nutrients requirement (i.e. protein, minerals, and vitamins) remain the same. Provision of dietary fat up to five to six percent will reduce the heat increment and also offer more metabolic water, which will have a cooling effect during the summer stress apart from improved palatability of feed. The feeding strategies of chicken layer during summer months are as follows:

1. Ensure that the non-energy nutrients, such as amino acids, vitamins, and minerals are increased in the formula in proportion to the decrease in feed intake.
2. When the nutrient density of the formula is increased to compensate for the reduction of feed intake, the protein content of the feed may be reduced by about 0.5 percent than the standard. At the same time intake of the essential and critical amino acids (lysine, methionine and threonine) can be optimized by providing enough quantities of synthetic amino acid preparations.
3. Avoid feeding the birds about three hours before ambient temperature is expected to be high in the range of 36°C (99°F) especially during midday. Adjust the lighting schedule to encourage the feed consumption in the night and early morning. Midnight feeding or an intermittent lighting program can encourage feed consumption at night.
4. Inclusion of Vitamin A 12,000 IU and Vitamin E -150 mg per kg diet for better performance and combat heat stress.
5. Vitamin C in the ration (150-300 gm per ton of feed) can protect birds from the effects of heat stress and enhance the survival of birds exposed to acute heat stress.

6. Use vitamin and electrolyte supplements in the drinking water. Shifts in the acid/base balance in the blood of heat stressed birds cause the loss of sodium, chloride, potassium, and bicarbonate in the urine.
7. During periods of high temperature, the flock has higher demand for drinking water. The water to feed ratio is normally 2:1 at 21°C (70°F), but increases to 4:1 at 40°C (104°F). For floor raised flocks, providing additional drinkers can help to fulfill the increased requirement.

Activity

1. Collect the common poultry feed ingredients in your locality with samples and document their nutrient profile.
2. Prepare a brooder, grower and layer feed formula with locally available feed ingredients.
3. Visit a commercial poultry feed mill and document the various processes in feed milling.

6

Chicken Egg Quality

Learning Objectives

1. Elaborate the formation of chicken egg
2. Describe the nutritive value and quality of chicken egg
3. Discuss the quality standards of chicken egg

Physical Structure of an Egg

The average weight of chicken egg is about 58 gram. The four principal parts of an egg are the yolk, the white, the shell membranes, and the shell. The yolk consists four major parts namely *latebra*, germinal disc, yolk material, and yolk membrane otherwise called as *vitelline* membrane. The egg white is called as *albumen*, has four distinct layers. Firstly, *chalaziferous* layer, which is adjacent to the yolk, and chord like structure called as *chalazae*. It keeps the yolk in the centre of the egg and it accounts for three percent of the egg white. The inner thin albumen layer surrounds the chalaziferous layer and makes up to 17 percent of the white. The firm or thick layer of white, act as pocket that, holds the inner thin white and the yolk. It is adjacent to the egg shell membrane and accounts for nearly 57 percent of the white. The outer thin layer attached inside of the shell membranes and accounts for about 23 percent of the egg white.

The shell membranes are fibrous in nature and composed chiefly of a *keratin* like protein. There are two shell membranes namely inner and outer shell membrane. The egg shell is composed of three layers namely inner mammillary layer, middle spongy layer and outer *cuticle*. In a freshly laid egg, there will not be any *air cell*. Air cell is air filled space found between two shell membranes, usually at the broad end. Once, the freshly laid eggs starts to cool, air cell will form between shell membranes. Structure of chicken egg is presented in Figure 6.1.

Figure 6.1 Structure of chicken egg

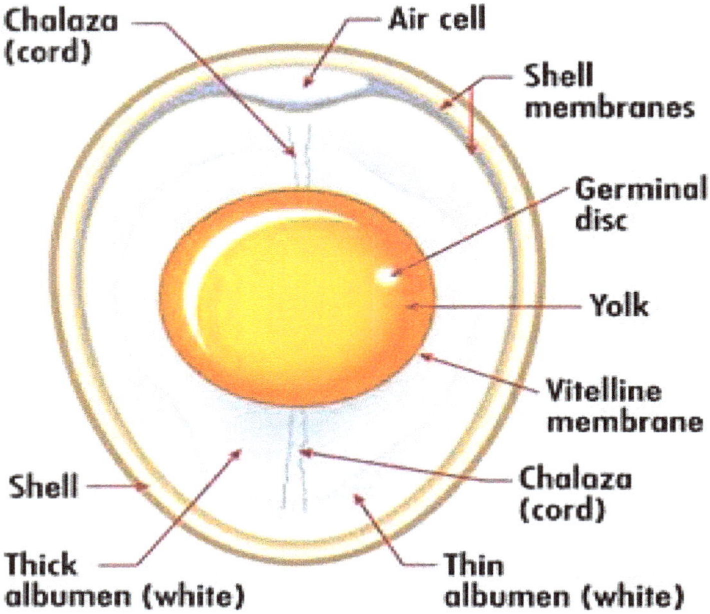

Composition of an Egg

Egg white (Albumen): The egg is a complete source of quality protein, fat, water and fat soluble vitamins and minerals. The complete protein portion of an egg is egg white or albumen. The fat content of egg white is very low. The principal component of albumen is water and it accounts of 88 per cent and protein 11 per cent. The pH of albumen of a fresh egg is between 7 and 8.5. Storage will raise the pH towards more alkaline due to loss of carbon di oxide. The major proteins of the egg white are *ovalbumin, ovomucin, conalbumin, ovoglobulin,* and *ovomucoid.* The albumen also contains water-soluble B vitamins. Riboflavin makes the greenish tint colour to the albumen.

Yolk: The two chief components of yolk are proteins and fats. The principal yolk proteins are *ovovitellin* and *ovolivetin.* The fatty substances of the yolk are triglycerides, phospholipids and cholesterol. The major mineral in the egg yolk is phosphorous. The pH of the yolk in a fresh egg is generally about 6.0 but it will increases to about 7.0 in storage. The popular yolk pigments are xanthophylls. Dietary yellow maize and green leafy plants are the source for yolk pigments. The yolk contains almost all the essential vitamins except Vitamin C. The vitelline membrane is water permeable and allows the inflow of water from egg white and plumbs the yolk and makes it less viscous during storage.

Table 6.1 Chemical composition of the chicken egg

Part of egg	Percent	Water	Protein	Fat	Ash
Whole egg	100	65.5	11.8	11.0	11.7
White	58	88.0	11.0	0.2	0.8
Yolk	31	48.0	17.5	32.5	2.0
Shell	11	94.0	1.0	1.0	4.0

Source : Egg - Grading Manual, USDA - 2000

Egg Formation

Female reproductive tract: The reproductive system of the hen consists of two parts namely ovary and oviduct. The unique feature of hen is that, left ovary and oviduct is functional in developing an egg and right is dormant. Ovary is a cluster of developing yolks, each separate from the others. It contains approximately 4,000 minute ova, each within its own sac or follicle. Oviduct is a long tube like structure present along the backbone and attached to it loosely. It is approximately 65 cm long. The oviduct is divided in five parts, namely, *infundibulum, magnum, isthmus, uterus* and *vagina* to perform various functions in the formation of egg and its laying.

Formation of the yolk: The yolk in the ovary starts its development as a single cell. It is called as female reproductive cell or germ with the surrounding vitelline membrane. The yolk slowly develops by the gradual addition of yolk fluid. It is layers of dark - light coloured yolk materials. Dietary fat soluble pigments, normally xanthophyll, are also transferred from blood stream to the yolk. The yolk matures as more yolk fluid is added. The germ stays at the surface of the yolk called as germ spot. A fast accumulation of yolk occurs 8 to 10 days before ovulation. This stage is triggered by the Follicle Stimulating Hormone (FSH) of the pituitary gland. The hormones secreted by the ovary in turn stimulate the oviduct to activity. The rate of yolk secretion increases rapidly from one week before ovulation and decreases thereafter. Ovulation in the hen is the release of a mature yolk contains ovum from the ovary. Here, ovum is the true female reproductive cell and not the egg yolk as a whole. Developing yolk is enclosed in a yolk sac or yolk membrane in which many blood vessels are present for supply of yolk materials. The area, which is free of blood vessels in yolk sac, is called as *stigma,* in which rupture of membrane for ovulation will take place. Occasionally, the rupture of the yolk sac occurs at sites other than the stigma. It may cause more blood vessels to rupture may excess bleeding result in blood spot in the yolk or white. Sometimes, brown, tan, or white spots, commonly known as meat spots, may occur in the egg. It is mainly due to discoloration of blood spots and debris of tissue sloughs in the reproductive tract. Generally ovulation repeats again about 30 minutes after an oviposition of a hen in its laying clutch. Act of egg laying in hen is called is called as oviposition.

Formation of the white: The first part of the oviduct is infundibulum, 7-10 long, is commonly called as funnel, because of its shape. When ovulation and release of yolk with ovum occurs, the funnel engulfs the yolk and directs its way to the oviduct. Apart, it also acts as a site for fertilization, if sperm present in the site. The yolk moves through the infundibulum to the magnum and the time taken for this movement is 15 minutes. Movement is brought about by peristaltic action. The magnum is the largest part of oviduct about 38 cm long and passage time for the yolk to travel through magnum is about 3 hours. Out of four layers of egg white, three layers are formed and almost all the protein in the white formed in the magnum. The white contains ovomucin, secreted by the magnum is fibrous in nature. It makes the thick white. These fibrous strands consolidate together as the developing egg making its spiral movement through the oviduct. From these strands, the chalaziferous layer and the chalazae are formed that is the first layer. The continuous twisting and drawing together of these strands tend to squeeze out thin white to form the inner thin white which is the second layer. The third layer is the remaining thick white is a homogeneous gel. The quality of the white is largely dependent on the amount of ovomucin secreted by the magnum part of the oviduct.

The isthmus is the next part of magnum, as the name implies, it is a constricted area of the oviduct. It is about 10 cm long, through which the developing egg passes for 80 minutes. Here, water and mineral salts are added and the two shell membranes are formed.

The uterus, next to isthmus is a part of the oviduct and it is also called as *shell gland* of hen. It is about 10 cm long and the developing egg stays here for the longest time, for about 21 hours for the formation of outer thin white and egg shell. In uterus, fourth layer of egg white, the outer thin white and minerals pass through the shell membranes by osmotic pressure, and the shell and shell pigment are added. The shell membranes are developed in a partly formed egg as it enters the isthmus and completed in uterus.

Formation of the shell: Egg shell composed of 94 per cent calcium carbonate. A hen may use as much as 50 percent of her skeletal calcium reserve for egg shell formation. Basically, there are two layers of the shell are formed in the *uterus*. The spongy layer consists of small calcite crystals that are not in any order, except in the outer portion of the layer, where crystals are set at right angles to the shell surface. Pigments are also deposited in the spongy layer of the shell and are derived from the blood. Pores are formed through the spongy layer connecting mammilla layer with the surface. Fully formed eggs are moving finally into the vagina, which is five cm long, then to *cloaca* and the *vent* for oviposition. The freshly laid egg has the pores and are filled by the matrix material and covered by the cuticle. An egg contains about 8000 pores. The passage time from ovulation to egg laying is slightly more than 24 hours. About 30 minutes after a hen has laid an egg, next release of yolk (ovulation) from ovary occurs for the next oviposition.

Abnormalities in Egg Formation

Sometimes, abnormal eggs will be laid by the hen due to factors like defective ovulation, yolk and shell formation and oviposition. The types of abnormal eggs are as follows,

Double-yolk eggs: When double yolks are released at a single ovulation and engulfed by the infundibulum, double yolk eggs will be laid by the hen.

Yolkless eggs: It is nothing egg without yolk. Sometimes, instead of an egg yolk, a bit of tissue that is sloughed off the ovary or oviduct, which may stimulates the secreting glands of the oviduct and a yolkless egg will be formed.

An egg within an egg: An egg is added to the next day's egg and the shell is formed around both. It is due to the reversal of direction of the egg by the wall of the oviduct.

Blood spots: They are caused by a rupture of one or more small blood vessels in the yolk follicle or rupture of blood vessels out of stigma area at the time of ovulation.

Meat spots: Meat spots in the egg have been formed mostly due to tissue sloughed off from the reproductive organs of the hen. Sometimes it may be due to blood spots which have changed their colour in the process of formation of egg.

Soft-shelled eggs (shell less egg): It generally occur, when an egg is prematurely laid, and insufficient passage time in the uterus prevents the deposition and formation of the shell and eggs will be laid without any shell.

Thin-shelled eggs: It may be caused by dietary deficiencies, heredity, or disease.

Glassy or chalky-shelled eggs: They are caused by malfunctions of the uterus of the laying bird.

Nutritive Value of Chicken Egg

Egg is natural, unadulterated and packed food, produced for chicken embryos for their nourishment during embryonic development by the mother hen. It contains all the nutrients like protein, fat, vitamins and minerals. The biological value of egg protein is high, which is considered as standard of comparison for other sources of protein like milk, fish and plant proteins. Biological value is a measure of protein quality considering the rate of efficiency of protein utilization for growth. The protein quality of egg is superior in terms of its amino acid makeup and it provides all the essential and non-essential amino acids. A large 50 gram egg contains 6.29 grams of high-quality protein, about 12.6 per cent of the Daily Reference Value (DRV) or Daily Values (DV) for protein for adults. These are the values that represent the amount of nutrient (protein, fat, carbohydrate, vitamins, minerals, and cholesterol etc.,) expressed in percentage of a specific nutrient that a person should consume per day.

Biological Values of Protein (On 100 point scale)	
Egg	93.7
Milk	84.5
Fish	76.0
Beef	74.3
Polished rice	64.0
Whole wheat	64.0
Soybeans	58.0

Source: FAO- United Nations. The Amino Acid Content of Foods and Biological Data on Proteins, Nutritional Study #24. Rome (1970)

Egg Candling

Candling is a method of inspecting eggs under focused light to determine the condition of the air cell, yolk, and white for any defects. Candling is done in a dark room with the egg held before a light. The light penetrates the egg and makes it possible to observe the inside of the egg. The instrument used for candling is called as candler. The candler requires minimum of 60 watt bulb for effective illumination. The candler should be set on a box or table at a convenient height (4 feet), so the light will not shine directly into the eyes of the operator.

Table 6.2 The nutritive value of chicken egg

Nutrient content of one large raw fresh egg (50 g)			
Nutrient	**Whole egg**	**Egg white**	**Egg yolk**
Energy (cal)	72	17	55
Protein (g)	6.3	3.6	2.7
Carbohydrate (g)	0.36	0.24	0.61
Total fat (g)	4.8	0.06	4.5
Monosaturated fat (g)	1.8	0.0	2.0
Polysaturated fat (g)	1.0	0.0	0.72
Saturated fat (g)	1.6	0.0	1.6
Cholesterol (mg)	186	0.0	184
Vitamin A (IU)	270	0.0	245
Vitamin D (IU)	41	0.0	37
Thiamin (mg)	0.02	0.0	0.03
Riboflavin(mg)	0.2	0.15	0.09
Vitamin B6 (mg)	0.09	0.0	0.06
Vitamin B12 (mcg)	0.45	0.03	0.33
Folate (mcg)	24	1	25
Choline (mg)	126	0.40	116
Vitamin E (mg)	0.5	0.0	0.44
Calcium (mg)	28	02	22
Phosphorous (mg)	99	05	66
Sodium (mg)	71	55	08
Potassium (mg)	69	54	19
Iron (mg)	0.88	0.03	0.46
Magnesium (mg)	06	04	01
Zinc (mg)	0.65	0.01	0.39
Selenium (mcg)	15.4	6.6	9.5

U.S. Department of Agriculture, Agricultural Research Service, 2010. USDA National Nutrient Database for Standard, Reference, Release 23. Nutrient Data Laboratory http://www.ars.usda.gov/nutrientdata.

In candling, the egg is held in a slanting position with the large end against the hole in the candler. The egg is grasped by the small end and, held between the thumb and tips of the first two fingers and turned quickly to the right or left. This moves the contents of the egg and throws the yolk nearer the shell. Brown eggs are more difficult to candle than white eggs. Candling is a skill and it requires basic knowledge about the egg contents and practice. One should be able to distinguish a fresh egg from a stale egg and detect abnormalities such as bloody whites, blood spots, meat spots, and cracked shells. In a fresh egg, the air space is plainly visible and moves freely and the white is thin and clear and yolk will give clear yolk shadow on the shell.

Nowadays commercial candlers are available with various dimensions. In developed countries, candling is a routine process for grading the shell eggs for consumers. Automatic candlers are available for this purpose. In India this is not the situation, market eggs are not being graded but marketed on number basis. However, manual candling is an effective method to teach and demonstrate the students of poultry science about the various components and conditions of the shell egg.

Quality Characteristics of Egg

In general, quality may be defined as the inherent properties of a product that determine its degree of excellence and level of acceptability. The conditions and characteristics that consumers want, and are willing to pay, are called as factors of quality. The quality of an egg is determined by comparing a number of factors. Quality factors may be classified in to two namely, exterior and interior quality factors.

Exterior quality factors: They are the factors that are apparent from direct external observation and should be the first point for evaluation.

Interior quality factors: The factors which involve the contents of the egg as they appear before a candling light, or when the eggs are broken out and measured by various methods plus visual examination of the yolk.

Exterior Quality Factors

The exterior quality factors of egg are egg weight, shape, air cell depth, shell colour & texture, cleanliness, volume, specific gravity and surface area.

Egg weight: The average egg weight of chicken is 58 gram. However, the range is 48-62 gram. Egg weight is influenced by many factors namely, breed, body weight of chicken, age of bird, selection history of the bird, protein content of the feed etc. As the age of the hen advances egg weight will increase. Egg weight can be recorded using digital balance with 0.1 gram accuracy. Though, in India, egg marketing is based on the numbers, AGMARK table egg weight standards are available for market eggs. Internationally, egg marketing is based on numbers and egg weights and USDA egg weight standards are also available for egg marketing.

Table 6.3 AGMARK chicken table egg weight classes

Sl.No.	Grade	Weight per egg (g)	Colour mark
1	Extra large	60 and above	White
2	Large	53-59	Red
3	Medium	45-52	Blue
4	Small	38-44	Yellow

Shape: The normal chicken egg has an oval shape with one end larger than the other, and it tapers toward the smaller end. The root of the word oval in Latin is *Ovalis*, means egg shape. Sometimes it is also called as Ovate. (Similar to oval). The ends of an egg are commonly called the broad end (air cell end) and the narrow end. Wide variation in shape may occur and they are called as abnormal shape eggs. Egg should have normal shape for easy packing, transportation and marketing. Shape of the egg is expressed as *Shape Index* and it is the ratio of breadth of egg to the length.

Shape Index = (Breadth of egg / Length of egg) x 100

Breadth and length of egg shall be measured using a *Vernier* caliper in centimeters. A normal egg will have a shape index of 72-74. Egg which is spherical in shape will have shape index values of more than 75. Those eggs which are elongated and elliptical will have lesser shape index of 70. Based on the degree of abnormality, eggs may be classified as practically normal and abnormal.

Air cell: Freshly laid egg will not have any air cell or may have a small air cell at broad end usually. As the egg cools, contractions of egg content will happen. It causes separation of inner and outer shell membrane leads to formation of air space called as air cell. It is also due to evaporation of moisture from the egg. The air cell is the easiest quality factor for evaluation. It can be measured using air cell gauge and depth of air cell is expressed in millimeters. The depth of the air cell is the distance from its top to its bottom when the egg is held air cell upward. In egg grading, the air cell depth is one of the factors in AGMARK and USDA standards. Free air cell is an air cell that moves freely towards the uppermost part in the egg as the egg is rotated slowly. Bubbly air cell is ruptured air cell resulting in one or more small, separate air bubbles, usually floating beneath the main air cell.

Shell colour and texture: Shell colour refers, whether shell colour is white or brown and shades of brown. Egg shell colour is mostly a breed or strain characteristic. Egg shell colour does not have any influence on nutritive value of the egg. Shell colour is due to the presence of pigments. *Ooporphyrin* gives brownish colour to the egg shell, which is normally seen in eggs laid by the Asian, English and American Class of breeds. Region based preference is prevails in various countries for different shell colour egg. In India, consumer preference is for white shelled eggs in commercial production and brown shell or tinted shell eggs in backyard poultry production. Shell colour is assessed by using a score card method and visual comparison. In this score, white shell eggs is having lowest score of 5 and based on the degree intensity of brown colour, score will increase up to 100. Shell texture refers the smoothness, roughness of shell surface and shell quality. Eggs that are unusual in shape may have ridges, rough areas, or thin spots on the shell surface. Abnormal shells may result due to nutritional deficiency or disease conditions Sometimes a shell is cracked while the egg is still in the body of the hen. These eggs, which are commonly referred to as *body checks*, they are repaired by an additional deposit of shell over the cracked area, generally resulting in a ridged area. Considering the quality of the egg shell, eggs may be classified as sound, checked and leaking.

Sound: An egg whose shell is unbroken.

Check: An individual egg that has a broken shell or a crack in the shell, but its shell membranes is intact and its contents do not leak.

Leaker: An individual egg that has a crack or breaks in the shell and shell membranes to the extent that the egg contents are leaking through the shell. Leakers are considered as loss.

Checks are unavoidable problem in marketing of eggs, because eggs cannot be assembled, graded, packed, transported, and marketed without some breakage. Such eggs could not be stored and should be diverted to immediate use or further processing. Body checks of egg may range from plainly visible dented cracks to very fine, hair line cracks otherwise called as *blind checks* that often escape in grading process. Blind checks are the most common, and frequently the most difficult to detect in rapid candling, it can be found out only by *belling*. Belling is the practice of gently tapping two eggs together detects the blind checks by sound and it needs considerable practice.

Cleanliness: This is essential for consumer satisfaction and also to improve and maintain the keeping quality. A dirty egg may harbour harmful microbes which will spoil the egg and make it unfit for consumption. Eggs collected from deep litter will be dirtier than caged eggs, due to dirty wet litter.

Volume: Volume is also one of the indicators of egg size. Egg volume is directly proportional to the egg size. Since, the specific gravity of egg is more than one, the volume of a fresh egg will always be less than the egg weight. To measure the volume of the egg, fill a measuring cylinder of one litre capacity with a known quantity of water. After noting the lower meniscus of the water, gently immerse the egg and slide the egg carefully into the measuring cylinder and observe the water level increase and note the final reading of the water, the difference in value will give the volume of egg. Volume of the egg will be expressed in cubic centimeters.

Specific gravity: This gives an indication of the egg shell quality, as well as its freshness. Fresh eggs will have higher specific gravity than stored eggs. There will be a loss of moisture in old eggs, which in turn replaced by air in air cell. So the air cell will become bigger as the egg is stored for a longer time. Similarly, eggs having stronger shell will have higher specific gravity than thin shelled eggs. The normal range of specific gravity of egg is 1.05-1.06. The measure of specific gravity in fresh can be done by the following method,

$$\textit{Specific gravity} = \textit{Weight of the egg (g) / Volume (cc)}$$

Surface area: Surface area of an egg is directly related to egg size. The elongated eggs will have more surface than spherical eggs. It could be calculated by the following formulae,

$$\textit{Surface area} = 12.6 \times ((\textit{Length (cm)} + \textit{Width (cm)}) \textit{ divide by } 4)^2$$

Where, 12.6 is a constant for chicken egg

Interior Quality Factors

The interior quality of egg can be measured by break open the egg and studying the various parameters of the egg like shell, albumen and yolk. The procedure for is to take a random sample of fresh eggs with normal shape, number the eggs and take the egg weight in gram. Candle the egg and record the air cell depth (mm) and break open the egg gently over the glass plate. Record height, width of albumen and yolk (mm) without disturbing the contents and record the shell weight separately. Remove the shell membrane and measure the shell thickness and record the defects if any such as blood and meat spots. Transfer the yolk into a petri-dish and record its weight and find out the weight of albumen by difference. Workout and record the albumen index, yolk index and Haugh Unit score and record the yolk colour score using *Roche* yolk colour fan

Albumen, yolk and shell: Albumen, yolk and egg shell per cent is calculated by considering each component weight to the total egg weight. In general mean albumen, yolk and egg shell constitutes 57, 32 and 11 per cent of its egg weight respectively.

Shell thickness: In a break open egg, remove the shell and peel off the membrane. Take three pieces of shell from three representative areas, narrow, broad and centre region. Record the mean thickness of shell by using Screw gauge and express in millimeter. Average range of shell thickness of chicken eggs is 0.33-0.34 millimeter.

Albumen index: The freshness and firmness of the egg white is correlated with the albumen quality. In a break open egg, the height of the thick albumen is measured using a tripod stand micrometer or a *spherometer*. The width of the thick albumen is measured by using the Vernier caliper. Albumen index is ratio of albumen height to the width. The height and width can be measured in three places to take the average values. A fresh egg will have an albumen index of 0.1. The height of the albumen will be high in a fresh egg and slowly lowers as the egg ages.

Albumen Index = Mean Height of albumen (mm) divide by
Mean Width of albumen (mm)

Haugh Unit Score (HUS)

Haugh Unit Score is the widely used measure for the albumen quality and freshness. It is a modified version of albumen index. Albumen height is influenced by the egg weight and size. Considering that, albumen height is adjusted to the egg weight for calculating Haugh Unit Score.

$$\text{Haugh Unit Score} = 100 \log (H + 7.57 - 1.7\, W^{0.37})$$

Where, W = Weight of egg in grams and H = Height of thick albumen in mm. A fresh egg will have a Haugh Unit Score of 90 and above. As the age of egg advances, the Haugh Unit Score will decrease.

Yolk index: This is a measure for standing quality of the yolk. It is the ratio of height of yolk to the width of the yolk. The height of the yolk is measured by micrometer and width by Vernier calipers. **Yolk Index = Height of Yolk (mm) divide by Width of Yolk (mm).** The average value for a fresh egg is 0.40. As the egg becomes aged, yolk flattened and the yolk index value is lowered.

Yolk colour: The range of yolk colour is light yellow to dark orange. Though, yolk colour does not have any influence on nutritive value of the egg, it has some consumer preference, people generally prefer dark yellow yolk. Yolk colour mainly influenced by feed ingredients. Diets which are rich in carotenoid pigments especially *xanthophill, cryptoxanthine, leutein* will impart dark yellowish / orange colour in the egg-yolk. Birds fed with diets rich in yellow maize, alfalfa meal, groundnut leaf meal will have a deep orange colour yolk. Yolk colour is assessed by the visual inspection using *Roche yolk colour fan*, which is having shades of yolk colour for comparison and score for the same.

Yolk: The quality of yolk can be assessed easily in candling and it is one of the good indicators of egg quality. The characteristics of the yolk are assessed by its shadow on the shell during candling. However, three yolk factors are considered in judging egg quality. They are distinctness of yolk shadow outline, size and shape of yolk, and defects and germ development.

Distinctness of yolk shadow outline: The distinctness of the yolk shadow outline is governed by three factors namely,

Thickness and consistency of the white: The thicker the egg white, the yolk shadow outline will be less distinct.

Condition of the yolk: This condition is determined by the presence or absence of defects that will be indicated as dark shadows on the yolk in candling light

Color of the yolk: It is difficult to determine the color of the yolk before the candling light. A deep colored yolk, would cast a darker shadow before the candling light than lighter yolk. The terms used to define the three degrees of distinctness of yolk shadow outline in the U.S. Standards of quality for shell eggs are,

Outline slightly defined: A yolk outline that is indistinctly indicated and appears to blend into the surrounding white as the egg is twisted.

Outline fairly well defined: A yolk outline that is discernible but not clearly outlined as the egg is twisted.

Outline plainly visible: A yolk outline that is clearly visible as a dark shadow when the egg is twisted.

Size and shape of yolk: The yolk of a fresh egg is round and firm. As the age advances, the strength of the yolk membrane weakens allowing water to be absorbed from the white. This increases yolk size and weight and causes stretching of yolk and weaken the vitelline membrane, is called as flattened yolk.

Defects and germ development: Yolk defects difficult to detect, unless they are prominent in nature especially in aged eggs or eggs with cloudy albumen. Germ development is visible before the candling light and viewed as a circular dark area near the center of the yolk shadow. If yolks are found with blood spot, egg will become inedible for quality reasons. In USDA egg quality standards. The terms used in defective egg yolk are,

Practically free from defects: A yolk that shows no germ development, but may show other very slight defects on its surface.

Serious defects: A yolk that shows well developed spots or areas and other serious defects.

Clearly visible germ development: Development of the germ spot on the yolk of a fertile egg, where it is plainly visible as a definite circular area.

Blood due to germ development: Blood caused by development of the germ in a fertile egg to a point where it is visible as definite lines or as a blood ring. It makes the egg inedible.

Egg white quality assessment: A fresh egg contains four layers of white namely, chalaziferous, inner thin, thick, and outer thin. The appearance of the egg during candling is governed largely by the relative proportions of the thick and outer thin layers of albumen. There are two important considerations about the albumen are included in USDA standards of quality of eggs. They are viscosity, and clarity of albumen. The condition of the albumen is determined in candling by the intensity of the yolk shadow and the freedom of movement of the yolk as the egg is twisted before the candling light. These factors are related to the viscosity of the albumen. Thick albumen permits only limited movement of the yolk, and an indistinct shadow results. The reverse is true of thin albumen, which permits free movement of the yolk, and a distinct shadow outline. In an egg break open study following are the terms used in USDA egg quality standards with regard to the quality of albumen. Generally the values of viscosity, Haugh unit score taken at temperature range of 45 and 60 °F (7.2 and 15.6 °C), and free from foreign bodies are taken as traits of quality.

Clear: Albumen that is free from discolorations or from any foreign bodies floating in it.

Firm: Albumen that is sufficiently thick or viscous to prevent the yolk outline from being more than slightly defined with a Haugh unit value of 72 or higher.

Reasonably firm: Albumen that is somewhat less thick or viscous than firm albumen with a Haugh unit value of up to 60.

Weak and watery : Albumen that is weak, thin, and generally lacking in viscosity, permits the plain visibility of yolk outline with a Haugh unit value less than 60.

Blood spots or meat spots: Small blood and meat spots with size of not more than one-eighth inch in diameter are to be classified as B quality. Eggs with larger blood spots or show diffusion of blood into the albumen are considered as loss.

Defects in Egg White and Yolk Conditions

Loss : Egg that is inedible, cooked, frozen, contaminated, sour, musty, or moldy, or an egg that contains a large blood spot, large meat spot, bloody white, green white, rot, stuck yolk, blood ring, embryo beyond the blood ring state, free yolk in the white, or other foreign material.

Bloody white: An egg which has blood diffused through the albumen. Eggs with bloody albumen are considered as loss.

Cooked eggs: Cooked eggs are eggs with coagulated contents and in candling presence of threadlike shadows in the white will be observed in slightly cooked eggs and in complete coagulation a dark, opaque appearance will be observed.

Musty eggs: Eggs appear clear and free from foreign material in candling light, and possessing musty odour is called as musty eggs.

Moldy eggs: Eggs with mold spots on the shell or mold growth in checked areas of the shell, or mold growths inside the egg itself is called as moldy eggs.

Large blood spots and bloody whites: They appear as brilliant red in color or as a dark gray in so-called meat spots, in contrast to the surrounding lemon-to-orange colored tinge of the yolk, observed before the candling light.

Green whites: This type of loss is caused by the *Pseudomonas* group of bacteria. Like sour eggs, eggs with green whites will fluoresce under the ultraviolet light when broken out.

Mixed rot: When the vitelline membrane of the yolk breaks and the yolk mixes with the white, resulting in foggy appearance throughout the interior of the egg when viewed before the candling light.

A white rot: It may be detected by the presence of threadlike shadows in the thin white with fruity odour.

Black rots: These eggs are opaque in candling light. When broken, the contents have a brown colour with a putrid odor and *Proteus* is main causative organism.

Stuck yolk: Stuck yolk occurs, when yolk membrane becomes attached to the shell membrane. It generally occurs in storage, the thick white becomes thin, the yolk floats close to the shell and becomes attached to the shell membrane.

Addled egg: In long time storage of eggs, the yolk membrane will lost its strength and breaks, mixing of egg yolk and white will occur, this condition is called as initially seeping yolk and later as addled egg.

Blood rings and embryo chicks: They are caused by germ development in fertile eggs by incubation temperatures. Aft the first day of incubation, the embryo develops a circulatory system, and death of embryo in this stage drains the blood to the outer edge of the germ disc and forms the blood ring. Under candling inspection, it appears as a brilliant blood red circle from one-eighth to three-eighths inch (0.3 to 0.9 cm) in diameter depending on the stage of development.

Egg Quality Standards

Indian Standards: In India, Bureau of Indian Standards (BIS) has prescription for the chicken eggs produced in India. They called as Table Eggs Grading and Marking Rules, 1968. Here, table egg means edible eggs derived as a product of chicken industry.

United States Standards: The United States standards, grades, and weight classes for individual shell eggs described by United State Department of Agriculture (USDA) are applicable only to eggs of the domesticated chicken that are in the shell.

Table 6.4 BIS grades of chicken table eggs

Grade	Weight / egg (gram)	Shell	Air cell	White	Yolk
A. Extra large A. Large A. Medium A. Small	60 and above 53 - 59 45 - 52 38 - 44	Clear unbroken and sound Shape is Normal	Up to 4 mm depth. Practically Regular	Clean reasonably firm	Fairly well centered Practically free from defects Outline indistinct
B. Extra large B. Large B. Medium B. Small	60 and above 53 - 59 45 - 52 38 - 44	Clean to moderately stained, sound and slightly abnormal	8 mm depth may be free and slightly bubbly	Clear may be slightly weak	May be slightly off centered outline slightly visible

Source : BI Standard grades of chicken table eggs, Govt. of India

Table 6.5 United States Egg weight classes for table shell eggs

Sizes of weight class	Egg weight gram per egg	Minimum weight per dozen of eggs (Ounces)
Jumbo	70	30
Extra large	63	27
Large	56	24
Medium	49	21
Small	42	18
Peewee	35	15

Source : USDA - Egg grading manual -2000, One ounce equal to 28.34 g

Table 6.6 USDA standards of quality of individual shell eggs

Quality factor	Specifications for each quality factor AA Quality	A Quality	B Quality
Shell	Clean, unbroken practically normal	Clean, unbroken practically normal	Clean to slightly stained* Unbroken, abnormal
Air cell	1/8 inch or less in depth. Unlimited movement and free or bubbly	3/16 inch or less depth, Unlimited movement and free or bubbly	Over 3/16 inch in depth. Unlimited movement and free or bubbly
White	Clear , Firm	Clear, Reasonably firm	Weak and watery. Small blood and meat spots present**
Yolk	Outlines slightly defined. Practically free from defects	Outline fairly well defined. Practically free from defects	Outline purely visible. Enlarged and flattened. Clear visible germ development but no blood. Other serious defects

For eggs with dirty or broken shells, the standards of quality provide two additional qualities.

Dirt	Check
Unbroken. Adhering dirt or foreign material, prominent stains, moderate stained areas in excess of B quality.	Broken or cracked shell, but membranes intact, not leaking***

* Moderately stained areas permitted (1/32 of surface if localized, or 1/16 if scattered.),

** If they are small (aggregating not more than 1/8 inch in diameter),

*** Leaker has broken or cracked shell membranes, and contents leaking or free to leak.

Table 6.7 Mean quality characteristics of fresh chicken egg

Sl.No.	Quality characteristic	Value
1.	Egg weight (g)	58
2.	Shape	Ovate
3.	Length (cm)	5.5 -5.7
4.	Width (cm)	4.1- 4.3
5.	Shape Index	72-74
6.	Surface area (sq.cm)	64-66
7.	Specific gravity	1.06
8.	Volume (cubic cm.)	50-52
9.	Air cell depth (mm)	3.2
10.	Egg white (%)	57
11.	Egg yolk (%)	32
12.	Egg shell (%)	11
13.	Albumen index	0.1
14.	Yolk index	0.4
15.	Haugh Unit Score	85-90
16.	Egg shell thickness (mm)	0.32 - 0.34
17.	Egg shell membranes thickness (mm)	0.006

Freshness of Eggs

Freshness of egg depends on number of factors and one of the critical factors is duration of storage. In storage, ambient temperature and humidity plays key role in deciding the shelf life of the egg. The ideal temperature humidity combination for storage of shell eggs is 2 - 4°C (35 - 40°F) and 70 - 80 per cent. In general, the recommended temperature and humidity range of domestic refrigerators are 4-7°C and above 70 per cent. Considering that, for domestic storage, it is desirable to store the eggs in refrigerator up to 3-4 weeks of time from the date laying without much quality reduction in freshness. When an egg is freshly laid, it's temperature will be around 41°C (105°F) with no air cell or very small cell. During storage as the advances, egg cools rapidly, the contents of egg contract. There inner shell membrane separates from the outer shell membrane to form the air cell. During aging, moisture and carbon dioxide leave through the pores of the shell, air enters to replace them and the air cell becomes larger. The separation of the shell membranes is the main reason for the easy peel of older eggs in boiled egg process. As the age of the egg advances, the egg white becomes thinner, the yolk becomes flatter and the yolk membrane becomes weaker. These quality changes don't have any serious effect on the nutritional quality of the egg or its functional cooking properties. However, refrigeration is much desirable for large scale pooling and storing of eggs. The eggs are having various functional properties namely binding, coagulative, emulsification and adhesive. Because of these properties eggs are being used in various food products like dairy products, confectioneries, cakes, soft drinks and food sausages.

Nutraceutical Properties of Egg: Traditionally, eggs are being used in China, India and European countries for many health benefits and other medicinal and industrial uses which are collectively called as special functional properties of eggs. The functional components of egg are as follows:

Lysozyme: It is an egg white protein has antimicrobial property used as a food preservative and as anti-microbial agent in pharmaceutical products.

Avidin : It is an egg white protein and has binding property with the vitamin biotin found in egg white and yolk. Avidin-biotin technology is currently being used in medical diagnostic applications such as immuno-assay, histopathological studies.

Sialic acid: It is an amino acid present in the egg has bacterial inhibition properties used in control of stomach infections.

Liposomes: They are fatty droplets found in eggs, are used as a controlled delivery mechanism for various drugs.

Immunoglobulin yolk: Globulin found in the yolk called as Immunoglobulin yolk (IGY), a simple yolk protein is used as an anti-human-rotavirus (HRV) antibody in food products.

Phosvitin: It is a phospho protein found in egg yolk, provides antioxidant benefits in food products.

Choline and Lecithin: Choline is a B vitamin combined with lecithin found in the egg yolk is crucial for brain development and is used to treat certain liver disorders. Eggs are one of the excellent food sources of choline.

Ovolecithin: It is a phospholipid present in egg yolk which has a high proportion of phatidycholine. It contains poly unsaturated fatty acids such as arachidonic acid (AA) and docosahexanoic acid (DHA) which are beneficial for human health.

Production System and Labeling of Eggs

Commercial chicken egg production system varies between countries, areas with in countries and between producers. In India, almost all the commercial chicken egg producers are adopting only cage system of rearing. However in developed countries, egg is being labeled differently based on the production especially housing systems. As per American Egg Board (AEB) the methods of labeling for marketing shell eggs is as follows, (*Source: The Incredible Edible Egg- Eggcyclopedia, Fifth Edition- American Egg Board, USA*)

Conventional: Eggs laid by hens living in cages with access to feed, water, and security. The cages serve as nesting space as well as for production efficiency. In this type of hen house, the birds are more readily protected from the elements of diseases and natural and unnatural predators.

Cage-free: Eggs laid by hens at indoor floor operations, sometimes called free-roaming. The hens may roam in a building, room or open area, usually in a barn or poultry house, and have unlimited access to fresh feed and water, while some may also forage for food if they are allowed outdoors. Cage-free systems vary and include barn-raised and free-range hens, both of which have shelter that helps protect against predators. Both types are produced under common handling and poultry care practices, which includes provision of floor space, nest space and perches

Free-range: Eggs produced by hens that have access to outdoors in accordance with weather, environmental or state laws. In addition to consuming a diet of grains, these hens may forage for wild plants and insects and are sometimes called pasture-fed hens. They are provided floor space, nesting space and perches.

Organic: Eggs produced according to national U.S. Department of Agriculture organic standards related to methods, practices and substances used in producing and handling crops, livestock and processed agricultural products. Organic eggs are produced by hens fed rations having ingredients that were grown without most conventional pesticides, fungicides, herbicides or commercial fertilizers.

Enriched Colony: A production system that contains adequate environmental enrichments to provide perch space, dust bathing or a scratch area (s), and nest space to allow the layers to exhibit inherent behavior. Enriched colony systems are American Humane Certified production system.

National Egg Coordination Committee

National Egg Coordination Committee (NECC) is an association of commercial chicken egg producers of India. It plays major role in market intervention, price support operations, egg promotion campaigns, consumer education, market research, rural market development and liaisons with the government on vital issues concerning the industry. It is a voluntary associative effort by producers. It is working principle is based on co-operative spirit to have, the right to fix their price for their products. The motto of NECC is *My egg, My price and My life.* It operates in three tier structure namely executive, zonal and local committee with the head office at Hyderabad. The objectives of the NECC are as follows, (*Source: www.e2necc.com*)

1. Price declaration.
2. To decide upon a reasonable price for eggs and to ensure profit for producers
3. To monitor the egg stock levels in different production centres.

4. Manage stock levels, regulate and balance demand and supply of eggs
5. Market intervention through Agrocorpex India Limited.
6. To organize and unite poultry farmers across the country.
7. To create a dependable distribution network
8. To generate employment by encouraging people to take up egg farming
9. To promote exports and develop export markets
10. To make available the technology and information
11. To get government support and financial aid from banks
12. Advertise with the objective of educating the customer.
13. Undertake a powerful egg promotion and consumption campaign
14. Market research, identification and its development.
15. Preparation and submission of industry position papers

Agricultural and Processed Food Products Export Development Authority (APEDA)

The Agricultural and Processed Food Products Export Development Authority (APEDA) was established by Agricultural and Processed Food Products Export Development Authority Act by Government of India. The head office is at New Delhi. The major functions of APEDA are as follows, (*Source: www.apeda.gov.in*)

1. Development of industries relating to the scheduled products for export by way of providing financial assistance
2. Registration of persons as exporters of the scheduled products
3. Fixing of standards and specifications for the scheduled products
4. Carrying out inspection of meat and meat products in slaughter houses, processing plants, storage premises, conveyances or other places
5. Improving of packaging and marketing of the scheduled products;
6. Promotion of export oriented production and development of the Scheduled products;
7. Collection of statistics from the owners of factories or establishments engaged in the production, processing, packaging, marketing or export of the scheduled products
8. Training in various aspects of the industries connected with the scheduled products;

APEDA is offering platform for exporting chicken shell eggs, whole egg powder, yolk and albumen powder, dressed poultry, cut up parts etc. to various countries.

Activity

1. Collect 24 eggs from a commercial layer farm and record the external and internal quality characteristics and classify the eggs based on AGMARK egg weight classes and offer your comments.

7

Bio-security in Commercial Layer Farm

Learning Objectives

1. List the critical steps in bio security programme of a poultry farm
2. Design a disinfection protocol for a commercial layer farm

Concept of Bio-security

The concept of bio security in a poultry farm facility is the collection of measures to be taken to prevent the entry, spread and attack of bio pathogens on birds. Basically, it works on three objectives,

1. To prevent the entry of disease producing organisms in to a farm
2. To prevent the spread of disease producing organisms within the farm
3. To minimize the incidence of zoonotic health problems
4. In a holistic bio security plan, there are three approaches and they are,

Conceptual bio-security: It involves selection of bio secure site for a poultry farm. The factors include virginity of soil for poultry farm operations, topographically ideal, logistically approachable, safe water, pollution free etc. In most of the cases, problems in conceptual bio security cannot be overcome at ease at later date; thereby it needs careful planning, discussion and decision making.

Structural bio-security: Poultry farm layout, construction buildings and sheds, distance between poultry sheds, feed mill , staff quarters , water tank, dead bird disposal plant etc., and its position related to others.

Operational bio-security: It consist of daily routine bio security managemental procedures like isolation of sick birds and quarantine of new stock, cleaning and disinfection of poultry houses, vaccination and medication, restriction of personnel entry, dead bird disposal- etc. This needs a detailed and well-structured schedule of programme for effective implementation.

General Routes of Pathogen Entry

1. Bird(new stock, transfers of birds, free ranging birds and sick birds)
2. Other animals (domestic animals, wild animals, rodents)
3. People (farm personnel, service personal, drivers, and visitors)
4. Equipment (equipment like feeders, drinkers and equipment etc.,)
5. Vehicles (feed trucks and egg lifting vehicle)
6. Air (aerosol transmission)
7. Water (contaminated water).

Strategies for Prevention of Pathogen Entry in to the Farm

1. Location of isolation of poultry farm facility
2. Peripheral fencing of the total farm area
3. Gate provision and gate log maintenance for visitors and vehicles
4. Provision of foot bath, truck wash, disinfection wash for humans
5. Procurement of birds from a disease free history flock
6. Complete restriction for pets and wild birds
7. Procuring clean feed ingredients or feed from the trusted sources
8. Systematic allocation of labours for different operations
9. Optimal down time of three weeks between batches for effective disinfection

The various bio security arrangements in commercial poultry farm is presented in Figure 7.1.

Figure 7.1 Bio-security arrangement in layer chicken farm

Security post and fence

Bio security sign post

Foot and truck disinfection bath

Disinfection shower bath for human entry

Bio-security Measures in Dead Bird Disposal

1. Dead birds will attract insect, wild and domesticated animals and act as a potential source of infection to the farm.
2. Collection and disposal of dead birds should be two times day and detailed post-mortem examination should be carried out before disposal.
3. Dead birds should be disposed properly by burial, incineration, composting or rendering based on the choice.
4. Carcass disposal trucks should be cleaned every day.

Disinfection in Poultry Farm Facility

Microbial contamination is a major cause factor for the incidence of disease. It can be prevented and controlled using proper set of management practices and that is called as disinfection and sanitation procedures and collectively called as decontamination process of poultry facilities. Decontamination process involves two steps namely *Cleaning* and *Disinfection*.

Cleaning: Cleaning is a process of physical removal of dust, soil, organic matter like droppings, blood and any waste secretions from a surface. Disinfection cannot be a substitute for poor cleaning and a good cleaning is a base for effective disinfection. Cleaning is of two types namely *dry cleaning* and *wet cleaning*. Dry Cleaning is the process of brushing, scraping, etc. of inside and outside the buildings, equipment etc. Wet cleaning involves cleaning with detergent mixed water like soaking, washing and rinsing of surfaces and equipment. Disinfection specifically denotes the use of chemicals in the cleaning process. Generally the list of chemicals used for disinfection process is based on their chemical nature, Alcohols (Ethyl alcohol and Isopropyl alcohol), Halogens (Idodophors, Chlorines), Quaternary Ammonium Compounds, Phenolic compounds, Aldehydes (Glutaraldehyde and Formalin), Oxidizing Agents (Hydrogen peroxide).

The choice of disinfectant for a particular process in based on number of factors and they are cleanliness of place, type of organisms in target, efficacy of the disinfectant against a microorganism, contact time, residual activity, cost and human health hazard problems. Effective implementation of disinfection procedures depends on right choice of disinfectant, for which clear understanding of different terminologies used in this process will be helpful.

Sterilization: The destruction of all infective and reproductive forms of all microorganisms (bacteria, fungi, virus, etc.).

Disinfection: The destruction of all vegetative forms of microorganisms. Spores are not destroyed.

Sanitation: The reduction of pathogenic organism numbers to a level at which they do not pose a disease threat to their host.

Formaldehyde Gas Fumigation

Fumigation with formaldehyde gas is an effective method for disinfecting the hatchery and all types of poultry houses. Formaldehyde is commercially available as formalin (37% formaldehyde in water). When the product is heated, formaldehyde gas is produced. It is quite toxic, with a tolerance of 5 ppm in air and inhalation should be avoided.

Directions for use of formalin: The heat necessary to release formaldehyde gas form formalin is produced by mixing potassium permanganate ($KMnO_4$) with it. Use an enamelware or earthenware vessel of large capacity because of the boiling, foaming, splattering action when the two chemicals are mixed. Place the vessel in the area to be fumigated, add the permanganate first in the vessel, and then add formalin. *Never add the permanganate to formalin.*

Two parts (by volume) of formalin are mixed with approximately one part (by weight) of potassium permanganate. This will produce complete expulsion of the gas. When the reaction is complete, a dry, brown powder will be left. If the residue is wet, not enough permanganate was used; if the residue is purple, too much permanganate was used. Varying concentrations of formaldehyde gas are necessary to fumigate under different conditions. The normal or single strength concentration (1 X concentration) is obtained by mixing 40 ml of formalin with 20 g of potassium permanganate for 100 cubic feet (2.83 cubic meters) of space. Likewise, 2 X, 3 X and 5 x concentrations are produced by respective increase in the quantity of chemicals for the same space. Empty poultry houses shall be fumigated with 3 X or 4 X concentrations for effective disinfection. Maximum disinfection efficiency could be obtained by maintaining the temperature of the area as 75 ^{0}F (24 ^{0}C), or higher, and the relative humidity as 75 per cent with the contact time of 20 minutes. In some countries formaldehyde fumigation is not a acceptable disinfection procedures due to human health risks.

Methods to Prevent the Incidence of Diseases

Vaccination: Vaccination is one of the important tools in disease prevention of a poultry flock. In intensive poultry production, almost birds are being protected through vaccination program from all the viral diseases. It is not possible to recommend standard vaccination program for all parts of country due to number of factors like, type of bird, endemicity of a disease, climate etc.

However, location specific and strategically sound vaccination programme will be helpful to prevent disease outbreak in most of the times, but, faulty vaccination will always fail to protect the birds from disease attack and sometimes it may act as source for infection. The purpose of vaccination is to improve the bird's immunity against a particular disease causing agent, thereby preventing from its pathogenic effects. Vaccines are generally delicate products, having shorter shelf life in storage (mostly cold storage), improper handling will reduce its potency in their action. The important part of vaccination program is recommended vaccination schedule for a particular farm, which should be prepared by the experienced native poultry consultant based on their experience in his work area or the prescriptions of the veterinary institutions or breeders specifications or combination of all these. Moreover, any vaccination program should be carried out under the supervision of qualified veterinary consultant.

Types of Vaccine

Live vaccine: The active part of the vaccine is the live organism that causes the disease. As such, it is sometimes capable of inducing the disease in over dose or faulty use in birds that have not had previous exposure to that organism. Cross infection can also occur.

Attenuated vaccine: In this type of vaccine, the organism has been attenuated by special procedures during manufacture, so that, it has lost its ability to cause the serious form of the disease but, the vaccine still has the ability to trigger the immune system to produce antibodies.

Killed vaccine: In this type of vaccine the organism has been killed and is unable to cause the disease, although the ability to trigger the immune system remains.

Vaccine forms: Vaccines are available in two forms namely liquid vaccine – it is in fluid form ready to use and freeze dried vaccine in powder form.Freeze dried form vaccine has longer shelf life and it should be reconstituted with diluents before use.

Common Vaccination Methods in Poultry Farm

There are a number of methods of vaccination available, the choice of particular method depends on vaccine and its type. In some of the methods that, individual vaccination has to be carried out and in some other, mass vaccination can be followed.

Ocular Route: This method involves the vaccine being put into one of the bird's eyes by a dropper. The vaccine makes its way into the respiratory tract via the lacrimal duct. The vaccine is delivered through an eye dropper and care must be taken to ensure that the dropper delivers the recommended dose.

Nasal Route: This method involves introducing the vaccine into the birds' nostrils either as a dust or as a drop and closing of one nasal orifice will make the bird to suck the vaccine inside immediately.

Drinking water: In this method, vaccine is mixed with water and administered as regular drinking water and it is convenient, stress free for birds and less time consuming. All equipment used for vaccination is carefully cleaned and free from detergents and disinfectants. Only cold, clean water of drinking quality should be used. Before vaccination, birds should be water starved for some time to increase the thirst and proper intake of vaccination dose through water.

Subcutaneous injection: This method is similar intramuscular injection but vaccine is injected under the loose skin (back of the neck or bottom of the wing) and care must be taken to ensure that the vaccine is injected into the bird and not just into the feathers or fluff in the case of very young chickens.

Intramuscular injection: This method involves the use of a hypodermic needle to inject the vaccine into the muscle (breast / thigh muscle) of the bird and nowadays automatic vaccinators available for maintaining accuracy of dose and ease in management and it needs some amount of training. However, in commercial poultry farming operations, ocular and nasal route methods are widely employed for live vaccines along with drinking water method and intramuscular route for killed vaccines.

General Guidelines for Proper Vaccination Programme

1. Procure vaccine from standard companies
2. Vaccination should be done at the cooler parts of the day
3. Check the expiry date and read the instructions carefully on the label of the pack
4. Exact dose of administration (extra label dose should be avoided)
5. Vaccination should not be done during disease or stress conditions
6. Young chicks should be vaccinated in such a way to avoid the maternal antibody influence
7. Equipment used for vaccination should be cleaned and sterilized
8. Vaccination should be done by trained personnel under veterinarian supervision

General Guidelines for Live Viral Vaccines Programmes

1. Live vaccines should always be kept in a refrigerator or in a freezer
2. Reconstitute the vaccine just two hours before its use.
3. Reconstituted vaccine should not be kept for longer time
4. Cool the diluents to 4 – 6°C before making dilution
5. Take small amount of vaccine in a dropper, so as to utilize it within 20 minutes
6. Empty vaccine and vaccine containers should be incinerated or deep buried.
7. Vaccination should be one go programme without any time gaps

General Guidelines for Vaccination through Drinking Water

In intensive chicken egg production environments, taking individual bird for vaccination is a time consuming process and vaccination through drinking water is easy vaccination procedure. However, it is having some limitations that ensuring the dose take up by individual bird cannot be ascertained. Moreover, primary vaccinations cannot be done through drinking water and it is highly suitable for secondary (booster) vaccinations. Considering that, following are the list of precautions to be taken during drinking water vaccinations.

1. All drinkers, utensils and pipelines used for vaccinations should be cleaned with non-medicated water
2. Drinking water to be used for vaccine preparation should not contain any detergents or medicines that may kill the organisms in vaccine.
3. Water should be cooled by adding ice and skimmed milk powder may be used at five gram per litre of water as a stabilizer of the virus in the vaccine
4. Birds should be made thirsty by withdrawing water for about two hours before vaccination.
5. Vaccine should be mixed in a quantity of water which is of 10 -15 per cent of the total water intake per day and that it would be utilized by birds within two hours.
6. Ensure every bird is taking sufficient water there by taking prescribed dose of vaccine.
7. Birds should not be disturbed during the vaccination process.

Table 7.1 Generalized vaccination schedule for commercial layer chicken

Sl. No.	Age	Name of the Disease	Route of Administration
1.	0	Marek's Disease	S/C
2.	7th day	Newcastle Disease (ND- F strain)	I/O
3.	14th day	Infectious Bursal Disease	I/O
4.	21st day	Infectious Bursal Disease -Booster	I/O
5.	28th day	Newcastle Disease (ND-Lasota)	I/O
6.	35th day	Coryza	S/C
7.	42nd day	Fowl Pox	I/M
8.	64th day	Coryza -Booster	S/C
9.	70th day	Newcastle Disease (ND- R_2B)	I/M
10.	80th day	Infectious Bronchitis	D/W
11.	85th day	Fowl Pox (Booster)	I/M
12.	105th day	Newcastle Disease (ND-Lasota)	D/W
13.	120th day	Newcastle Disease plus Infectious Bursal Disease	S/C
14.	36th week	Newcastle Disease (Lentogenic)	D/W
15.	40th week	Newcastle Disease plus Infectious Bursal Disease	S/C
16.	52nd week	Newcastle Disease (Lentogenic)	D/W
17.	65th week	Newcastle Disease (Lentogenic)	D/W

S/C- Subcutaneous, D/W-Through drinking water, I/O- Ocular/Nasal route, I/M-Intramuscular

Methods of Poultry Carcass Disposal

Dead birds of a poultry farm are potential source of pathogens for causing disease spread with in the farm and outside the farm. Considering this problem, utmost care should be taken to dispose the dead birds effectively. Collection of carcass should be carried out twice a day and on the day itself it should be disposed scientifically after thorough post mortem examination. There are number of ways to dispose the carcass and each method is having merits and demerits. Considering the size of operation, cost and its effectiveness, suitable methods can be selected and employed.

Burial in pits: Burial pits are simple, convenient, inexpensive disposal method. Burial pit site should be minimum of 60 meter away from, poultry sheds, drinking water sources and human residences. Well designed disposal pit of 5 x 2 x 3 m (30 m^3) should be sufficient to handle carcasses from a 10,000 chicken farm for two years, assuming a normal mortality of 5-7 per cent per batch.

Burning: Burning is the most effective and complete way of destroying infectious material including carcass. Nowadays smokeless and odourless incinerators are available for the disposal of chicken carcasses. Position of incinerators should be placed downwind direction of both poultry houses and human residences. Carcasses must be reduced to a completely burnt white ash. Capital investment, fuel cost and existing regulations in a particular locality are the main considerations in this system.

Rendering: Rendering is the most scientific method of converting chicken carcasses into a valuable, biologically-safe protein byproduct meal. Rendering is defined as a process of using high temperature and pressure to convert whole poultry carcasses or their by-products in to safe, nutritional and economically valuable products. Transport of chicken carcass to rendering facility is an additional process in large scale operations and cost intensive. Bio security of rendering plant site is critical in large scale operations. The rendering temperature of 130-140^0 C for 20-30 minutes is sufficient for disinfection and even sterilization of the rendered product. It is highly suitable for large scale operations with centralized rendering and carcass disposal process.

Compost: Composting is a controlled, natural process in which beneficial microorganisms of bacteria and fungi reduce and transform organic waste into a useful end-product. Composting chicken carcasses provides an economical and biologically safe means of converting carcass into an odourless, humus-like material which is useful as manure for soil fertility and a source of plant nutrients. However the use of composting in commercial poultry operations is very minimal because of its inherent complex process plan.

Activity

1. Prepare a bio security plan for commercial layer farm with suitable disinfection procedures and vaccination programme.
2. Collect the data on the disinfectants available for poultry farm from market and offer your comments.
3. Collect the data on the vaccines available for commercial layer chicken from market and offer your comments.
4. Visit a commercial layer farm on a vaccination day to get the practical experience on vaccination.

8

Common Poultry Diseases

Learning Objectives

1. List out the viral, bacterial , fungal and metabolic diseases of poultry
2. Discuss the signs, lesions, treatment and prevention of poultry diseases

Introduction

In commercial poultry production environments, the problems of disease occurrence and its economic losses posing great challenges to poultry farmers. Generally, when the normal body functions are getting impairment, disease occurs and degree of impairment determines the degree of severity of the disease. Occurrence of diseases has multi-faceted etiology and factors may be classified in to viral, bacterial, parasitic, nutritional etc., Diseases caused by infectious and parasitic agents are mostly complex in nature, which involves host, agent and environment. Nutritional deficiency induced conditions are mostly temporary and reversible in nature.

Viral Diseases

Newcastle Disease (ND): Newcastle disease, otherwise known as Ranikhet disease in India is the most common contagious disease in commercial chicken production caused by *Paramyxovirus*. New Castle disease virus has three strains based on virulence namely, mild (*lentogenic*), medium (*mesogenic*), and virulent (*velogenic*). Transmission is through air, infected materials, birds and equipment. The incubation period is about 3 to 6 days. The clinical signs in respiratory and nervous form are respiratory distress, labored breathing with wheezing, accompanied by nervous signs, such as paralysis or twisted necks called as *torticollis*. In intestinal forms the lesions are pin point hemorrhagic lesions in the proventriculus and hemorrhages all through intestine, diarrhea and inflamed tracheas. Affected birds will have drop in egg production up to 50 per cent and may return to normal production 2-3 weeks after the subside and, it will also affect the egg quality. Diagnosis is done by clinical signs followed by laboratory confirmation. Confirmation can be obtained with virus isolation and identification from tracheal or cloacal swabs and serology testing with haemagglutination tests. There is no specific treatment for this disease and use of antibiotics will reduce the secondary bacterial infections. Prevention programme is very effective in commercial layers and birds should be vaccinated with suitable vaccination programs with live, attenuated and killed vaccines.

Infectious Bronchitis (IB): Infectious Bronchitisis is an acute, contagious, and economically important disease. It is caused by an *Avian Coronavirus*. The transmission of infection is through bird to bird and air borne route. The incubation period is only 1-3 days. The prominent clinical signs in young chicken are the respiratory form with gasping, sneezing, tracheal rales and nasal discharge with depression and off feed. Generally mortality will be low, however secondary bacterial infections will aggravate the condition and increase the mortality. In adult layers and breeders, combined with initial respiratory signs, the flocks will have a severe drop in egg production and egg quality with defective egg shell and internal egg quality changes. Diagnosis of this disease is through clinical signs and post mortem lesions in a flock followed by laboratory confirmation based on virus isolation and identification from the infected samples. There is no specific treatment for this disease and antibiotics are used to control secondary bacterial infections. Vaccination with strain specific or cross protective live and inactivated vaccines at point of lay will protect the birds from the disease occurrence.

Infectious Bursal Disease (IBD, Gumboro disease): The disease is caused by *Birnavirus*. The virus is considerably stable and difficult to eradicate from an infected farm. Transmission of IBD virus is through, direct contact, infected droppings feed, water and equipment. The age of susceptibility is 3-8 weeks. Clinical signs are listless, depression, huddling and watery white diarrhea. Mortality may vary from 10-50 per cent depending on the pathogenicity of the strain involved. IBD infection in early ages will result in permanent immunosuppression without mortality and immunosuppression is economically important due to increased susceptibility to secondary bacterial infections especially in the respiratory tract. Diagnosis is through post mortem lesions like, in acute cases the *Bursa of fabricius* is enlarged and gelatinous, sometimes even bloody, haemorrhages in breast and thigh muscles and pale kidneys. The lack of white blood cells results in a reduction in the development of immunity and decreased resistance of the birds to other infections. Histopathological examination, serology, virus isolation are confirming diagnostic procedures. No effective treatment available for this disease. Prevention is through vaccination of layer birds.

Avian Influenza (AI): Avian Influenza is caused by an *Orthomyxovirus*, having several serotypes. As per current status, there are 16 H- types and 9 N-types and their combinations. The most important serotypes are H5, H7 and H9. The virus is excreted from nostrils, mouth, conjunctiva and cloaca. Airborne virus transmission, contact with infected droppings, people, cloths and equipment are the main source of infection. Infected water fowl and migratory wild birds may also act as carrier for the source of infection. In commercial poultry production, main disease occurrence happening in chickens, turkeys and ducks. Clinical signs will vary, depending on the pathogenicity (High Pathogenic AI and Low Pathogenic AI) of AI virus involved and other factors as host species, sex, concurrent infections, immunity and environmental factors etc., Low Pathogenic AI shows generally mild symptoms like depression, off feed, coughing sneezing, lacrimal secretions, nasal discharge, and limited drop in egg production and ends in low mortality rate. High Pathogenic AI produces fast onset with more acute symptoms like deep depression, drop in feed and water intake, severe drop in egg production and causes mortality of 50-90 per cent. Though, clinical signs are indicative for the presence of disease, confirmatory diagnosis by laboratory testing is mandatory, because of its possibility of zoonotic importance. Direct detection of AI proteins or nucleic acids from infected organs, tracheal or cloacal swabs and serology of blood samples for routine detection of antibodies. There is no specific treatment for Avian Influenza like any other viral diseases. Antibiotics will help to prevent secondary bacterial infections. In many countries AI is a notifiable disease with specific local regulations on its control. In AI free areas, the disease is controlled by complete culling and disposal of affected birds in specific area of coverage.

Infectious Laryngotracheitis (ILT): Infectious Laryngotracheitis is caused by *Herpes virus.* The entry of virus is through upper respiratory tract and ocular route. It is also spread through direct contact from bird to bird and by contaminated people or equipment. Incubation period varies from 4 -12 days. The main species affected is chicken. Clinical signs are acute respiratory symptoms with nasal discharge and rales, gasping, respiratory distress and expectoration of blood-stained mucus in birds. Egg production can drop up to 50 per cent, but will become normal after three weeks. The range of mortality is from 10-60 per cent. Lesions are found throughout the respiratory tract and more significant in larynx and trachea with classical tracheitis with hemorrhagic and diphteric changes. Diagnosis is through clinical signs and confirmatory diagnosis by demonstrating intra nuclear inclusion bodies in tracheal epithelial cells in histopathology and virus isolation. There is no specific treatment for this disease and prevention is through suitable vaccination.

Fowl Pox (Avian Pox, Avian Diphtheria): Fowl pox is caused by *Poxvirus.* Transmission is through direct contact and contaminated feed, water and equipment. Mosquitoes and other insects can also transmit the virus. The incubation period varies from 5 to 15 days. The major species affected is chickens, and pigeons by different fowl Poxvirus strains. Clinical signs are mainly external, mainly on the head or internal as wet pox affecting oral cavity, esophagus and trachea with pox lesions. Pox lesions can also be found on other parts of the body like skin of legs, cloaca etc. The lesions on the head, combs, and wattles are usually wart-like in appearance, yellow to dark brown in color. The internal lesions in the mouth, oesophagus and trachea are yellow-white and cheesy in appearance. Affected birds will be depressed and off feed, drop in egg production and in wet pox conditions labored breathing. Mortality is variable and depending on the severity of infection it may go up to 40 per cent. Diagnosis is observing wart-like lesions of the head mainly on comb, around the eyes or wet pox lesions of the mucous membranes of the nasal and oral cavities. A definitive diagnosis can be made in a laboratory by histological examination (inclusion bodies) or virus isolation in embryonated chicken eggs. There is no effective treatment. Control is preventive vaccination using a live vaccine is a successful method. In fowl pox disease outbreaks, it is advisable to vaccinate the flock immediately, which is called as emergency vaccination to stop further spread of the disease.

Lymphoid Leucosis (Big liver disease, Visceral leucosis): Lymphoid leucosis is caused by *Retrovirus*. Transmission is vertical (through eggs) and horizontal. Clinical signs are tumors in the visceral organs like liver, spleen, kidneys, and bursa of birds. Affected birds will have a reduction in egg production. Diagnosis of the infected birds is through testing blood samples and virus isolation from eggs and cloacal swabs. The tumors in the lymphoid leucosis may cause difficulty in differentiating from Marek's disease, but in Lymphoid leucosis nervous system will not be affected. Histopathological examination is essential for a proper diagnosis. The tumours caused by this virus are normally seen from sexual maturity onwards and are frequently located on the surface of bones such as the junction of the ribs, sternum, pelvis, mandible and skull and may be also found in visceral organs. No effective treatment is known. The best control method is the detection of infected birds through laboratory tests. Eradication of the virus from primary breeding stock through culling is the most effective way of controlling infections.

Marek's Disease (Neurolymphomatosis): Marek's disease is caused by *Herpes virus*. It is highly contagious in nature and transmission is by infected materials and premises, and chicks will get infected by the oral and respiratory routes. Feather follicles of infected chickens can act as a potential source of infection. The clinical signs are weight loss or some form of paralysis. The classical sign is neurolymphomatosis with involvement of leg nerve causes a bird to lie on its side with one leg stretched forward and the other backward. Subclinical infections result in impaired immune responses as disease causes a lytic infection in lymphocytes. Mortality usually occurs between 10 and 20 weeks of age and can reach up to 50 per cent in unvaccinated flocks. Diagnosis is through the presence of tumors in liver, spleen, kidneys, lungs, ovary, muscles, or other tissues. Skin involvement often consists of tumors of feather follicles or in between follicles. A proper diagnosis to differentiate Marek's from Lymphoid Leucosis requires specific histological study and the paralysis is caused by lesions and enlargements of the affected nerves. Prevention is through vaccination and MD vaccine only prevents the formation of tumors and paralysis. It is therefore of major importance to maintain high hygienic and sanitary measures to prevent the exposure. Vaccination against Marek's disease is performed in the hatchery itself in day-old chicks.

Avian Encephalomyelitis (Epidemic tremor): Avian Encephalomyelitis is caused by an entero virus belonging to the *Picornavirus* group. Egg transmission is the major route of transmission of this virus. Infected birds will show a drop in egg production. Infected chickens that hatch will show clinical signs of the disease and spread the infection in the incubator to other chicks. The usual clinical signs are tremors, increased mortality in young chickens, together with a drop in production in layers. Affected chickens sit on their hocks, difficulty in movement and will fall on their sides. A mild, trembling of the head and neck can be observed in young chick. Mortality in affected flock may go up to 50 percent. Diagnosis is through clinical signs like tremors, drop in egg production and hatchability in laying birds. Differential diagnosis should be made for Vitamin E deficiency. Demonstration of antibodies in the serum or hatching eggs will confirm the disease. There is no effective treatment for Avian Encephalomyelitis. Prevention is through vaccination of breeder pullets and layer pullets with suitable vaccination program.

Chicken Anaemia (Blue wing disease): The causative agent for chicken anaemia is *Circoviridae*. The principal route of transmission is vertical, from infected breeder hens. Horizontal transmission from bird to bird or by infected equipment, clothing, etc. is also possible. Adult birds may get infected, but will not develop clinical signs. The disease is characterized by depressed birds, increased mortality and anaemia and retarded growth. Lesions are thymus atrophy, bone marrow atrophy, subcutaneous and intramuscular hemorrhages with atrophy of the lymphoid system. Affected birds may show focal skin lesions often infected by bacterial infection, which also known as blue wing disease. Mortality rates vary, but generally about 30 per cent. Diagnosis can be based on the clinical signs and pathological findings in affected birds. Virus isolation is also possible. There is no specific and effective treatment for chicken infectious anaemia. Prevention is through vaccination of breeders with live attenuated vaccine for uniform maternal antibodies to prevent vertical transmission.

Egg Drop Syndrome (EDS 76): The disease is caused by a virus belonging *Avian Adenovirus*. Transmission of virus is through the egg to a few birds in a flock, these carrier birds at the point of lay of flock, begin to excrete virus and infect birds kept in the same house. Horizontal spread through infected litter can also occur once a flock shows the disease condition. The main species affected is chickens. Clinical signs of the disease are anaemic, diarrhea, reduced feed intake and failure of the layers to reach peak egg production or a drop in egg production in upswing production period, accompanied by an inferior egg shell quality and a loss of shell color in brown shell eggs. Morality is very minimal and no specific internal lesions will be available for diagnosis. There is no effective treatment against egg drop syndrome. Prevention is through vaccination with an inactivated vaccine before point of lay.

Bacterial Diseases

Infectious Coryza: Coryza is a bacterial disease caused by *Avibacterium paragallinarum, Haemophilus paragallinarum*).There are three common serotypes namely A, B, C are present in this organism. The disease spreads through airborne route, direct contact and through contaminated feed and water. The incubation period varies from 1 to 3 days. The principal clinical signs are acute inflammation around the eyes and upper respiratory tract, serous to mucoid discharge in the nasal passage and sinuses, facial edema and conjunctivitis. Feed and water consumption will be decreased resulting in loss of weight gain and loss of egg production up to 50 per cent in layers. Mortality will vary with the virulence of the infection and complicating infections with other viral and bacterial infections will cause substantial mortality. Diagnosis of this disease is through clinical signs, symptoms and confirmatory diagnosis is done by the isolation of the organism from the sinus or air sac exudate of the affected birds. Eradication is not economically feasible. Treatment with various antibiotics will alleviate the severity and course of the disease. Vaccination with multivalent serotypes is the preventive method and it is preferred in endemic areas.

Colibacillosis: The Colibacillosis is caused by avian pathogenic *Escherichia coli* bacteria. Infection can be of primary or secondary type. Transmission is through direct contact with infected feed, water, droppings and insects etc., Clinical signs are depression, off feed, lameness, and mortality. Lesions are colisepticaemia in combinations with poly serositis, salpingitis, osteomyelitis, synovitis, enteritis, meningitis and young birds with omphalitis. Both morbidity and mortality are variable. Diagnosis is through clinical signs in combination with isolation and identification of the organism. On the treatment side, antibiotics can be used based on antibiotic sensitivity test. Prevention is through strict hygiene management through bio security measures in the farm.

Chronic Respiratory Disease (CRD): The causative agent of Chronic Respiratory Disease is *Mycoplasma gallisepticum*. One of the triggering factor for onset of Chronic Respiratory Disease, is any disease like New castle disease and complication with secondary bacterial invasion. The two main organisms involved in this complex infection are *Mycoplasma gallisepticum* and *Escherichia coli.* Stress induced by transport, animal handling for beak trimming, weighment, and other unfavorable conditions like extreme weather conditions, are predisposing factors for the occurrence of this disease. This disease is having vertical and horizontal transmission routes. In vertical transmission, breeder birds can act as source of infection for chicks. Airborne, direct and indirect contact transmission are the routes for horizontal transmission. The incubation period varies from four days to three weeks. The prominent clinical signs are respiratory distress, sneezing, lack of appetite, decreased weight gain and egg production in layers. Post mortem lesions are reddish inflamed trachea, cheesy exudates in the air sacs in complicated cases especially, when it is combined with *Escherichia coli* infection. Generally, the disease will be of chronic in nature, will not cause high mortality and however because of its chronic nature, affects production substantially. Diagnosis can be made based on clinical signs and postmortem lesions followed by confirmation in the laboratory tests. Treatment of this disease is through suitable antibiotics.

Fowl Cholera (Avian Cholera, Pasteurellosis, Avian hemorrhagic septicaemia): Fowl cholera is caused by bacteria called *Pasteurella multocida.* Transmission of fowl cholera is mainly from bird to bird by water or feed contamination. Vectors like flies and red mite can be carriers. Rodents also play a role in contamination of water and feed with *Pasteurella multocida*. Clinical signs are depression, loss of appetite, bluish comb and swollen wattles. Egg production will drop 5 to 15 per cent and mortality will be high in acute fowl cholera. Lesions are acute phase septicaemia, vascular changes in abdominal viscera, hemorrhages, liver swelling with focal necrosis, ovaries appear flaccid and hemorrhagic and ruptured yolks. In chronic phase, localized infections in conjunctiva, facial edema, middle ear infection resulting in torticollis and meningeal infection. Diagnosis is through clinical signs in combination with isolation and identification from infected samples from fresh dead birds. In treatment side, antibiotics can be used based on antibiotic sensitivity test. Prevention is through hygienic management and rodent control to eliminate possible sources of infection. Vaccination can be considered in areas where occurrence is prevalent with both live and inactivated vaccines.

Pullorum disease and Fowl Typhoid: Pullorum disease is caused by a bacteria name by *Salmonella pullorum* and fowl typhoid is caused by *Salmonella gallinarum*, both are related, but not identical. Pullorum and typhoid can be transmitted vertically by infected carrier breeder hens through their eggs and horizontally. Chicks hatch out of infected eggs will have pullorum disease with early chick mortality. Infected chickens can also infect other chicks via droppings. However fowl typhoid is a disease mostly affecting adult chickens, with high mortality and morbidity. Horizontal transmission is common with fowl typhoid through infected droppings, dead birds, clothing, equipment and other fomites. Clinical signs in young chicks hatched out of *pullorum* or *gallinarum* infected eggs will show weakness, depressed appetite, poor growth and increased brooder house mortality. In older birds, depression, diarrhea, ruffled feathers, pale shrunken combs and drop in egg production and mortality will be observed. Morbidity and mortality can be highly variable and can reach 60 per cent. Lesions includes enlarged and congested liver, spleen and kidneys, pericarditis and hemorrhages in acute form. In adult birds, atrophy of ovarian follicles may be found. Treatment is through use of organism sensitive antibiotics, which will not completely cure, but will reduce mortality. The classical way of control is elimination of infected carrier breeder hens. Blood testing by the serum plate agglutination test with suitable *pullorum* antigen will detect carrier birds for culling. In commercial layer operations, vaccination against fowl typhoid is also in practice.

Mycoplasma Synoviae infection (MS): *Mycoplasma synoviae* infection frequently occurs as subclinical upper respiratory tract infection involving air sac lesions. In systemic infection of *Mycoplasma synoviae*, it can cause acute or chronic infection of synovial membranes of joints and tendons resulting in synovitis, or bursitis. Transmission is through vertical and horizontal routes. Prominent clinical signs are pale comb, lameness, retarded growth and ruffled feathers, swelling of joints and breast blisters in severe cases. Mortality will be very low and in layer it may cause drop in egg production and misshapen eggs. Post mortem lesions are synovitis, a viscous creamy to grey exudate involving synovial membranes of tendon sheaths, joints and bursa. Diagnosis is through serology and isolation and identification of organism. Treatment is through suitable antibiotics and prevention is through vaccination.

Fungal Diseases

Aspergillosis (Brooder Pneumonia / Fungal Pneumonia): The principal fungus causing Aspergillosis in poultry is *Aspergillus fumigatus*. Transmission is through inhalation of spores from contaminated deep litter and feed. Young chickens are very susceptible. Clinical signs are depression, thirst, respiratory symptoms like gasping and rapid breathing. Mortality will go up to 50 per cent. Since, this disease affecting respiratory system chiefly, gross lesion involves lungs and air sacs primarily and yellowish white pin head sized lesions in these organs. Occasionally, all body cavities are filled with small yellowish green granular fungus growth. Diagnosis of this condition is through identification of the presence of *Aspergillus fumigatus* organisms through microscopical study or even with the naked eye in the air passages of the lungs, air sacs or in lesions of the abdominal cavity. They may be confirmed by isolation and identification of the fungus from lesions. There is no specific treatment for infected birds. The first strategy is to cull the affected birds. Hygienic and quality litter and feed management are essential to prevent the infection.

Coccidiosis: Coccidiosis in chicken is caused by parasites of the genus *Eimeria,* which includes the five economically critical species namely *Eimeria acervulina, Eimeria maxima, Eimeria. tenella, Eimeria necatrix and Eimeria. brunetti.* Transmission is through feces containing sporulated *Eimeria* oocysts. The incubation period from ingestion to shedding of new oocysts is four to seven days. Clinical signs are weight loss, bloody droppings and vary by the species infecting the flock. *Eimeria tenella* and *Eimeria necatrix* can induce sick and listless birds in combination with high mortality. Mortality in field cases of these species will be high, if it is complicated with secondary bacterial infections like *Clostridium perfringens.* Diagnosis is through clinical signs and post-mortem examination, including intestinal mucosal scrapings of affected birds. Proper diagnosis requires microscopic examination of scrapings taken from the mucosa of the middle intestine. Hemorrhages of the cecal mucosa and blood filled ceca strongly suggestive of *Eimeria tenella* and confirmed by demonstration oocysts in mucosal scrapings study under histopathology. Swelling of the middle intestine with tiny white spots and petechiae on the serosal surface called as *salt and pepper* indicate *Eimeria necatrix* and may be confirmed by histopathology. In the part of treatment, anticoccidial chemicals are effective against *Eimeria* spp.however, longer duration of use will develop resistance against the drugs. One of the effective ways to develop controlled immunity is through feed medication with ionophore antibiotics like monensin. The ionophores reduce coccidiosis multiplication, allowing natural immunity to develop in the flocks. Another method to develop controlled immunity is coccidiosis vaccination with live attenuated sporulated oocyst vaccines.

Worm infection: Worms of chickens are classified into four categories and they are Roundworms (*Ascarid*), Hairworms (*Capillaria*), Caecal worms (*Heterakis*), Tape worms (*Raillietina and Davainea*). Transmission of round worms is through eggs shed in the faeces and tape worms are through shed of gravid proglotids carrying eggs. Clinical signs and lesions includes damage of intestinal lining, causing enteritis, anaemia, decreased egg production and at times eggs with pale yolks. *Capillaria* cause more damage to the intestinal lining and can cause enteritis and anaemia with decreased egg production and the appearance of pale egg yolks. Tape worms cause inefficient use of nutrients in chicken affecting economic performance. Diagnosis is through post mortem examination of the intestinal contents, which will reveal roundworms, caecal worms, and tape worms. Treatment is through use of broad spectrum of specific anthelmintics and prevention is through hygienic litter management programs. A worm burden leads to reduced immunity, poor feed efficiency. Therefore, deworming must be practiced regularly to maintain healthy flock. Deworming is mostly done through water, but is also done through feed. The first deworming is done at seven weeks of age, followed by at 14 to 16 weeks of age. During laying period, deworming should be done at least bimonthly in deep litter system of rearing and once in three months in cage rearing.

Metabolic Diseases

Gout: The gout is a condition characterized by buildup of urates in the system and deposited in tissues and called as two conditions namely, visceral gout and articular gout. The visceral gout is manifested with deposition of urates in renal tubules and the serous surface of the heart, liver, air sacs or the peritoneum looks like a chalky white dust. Visceral urate deposits are generally due to renal failure. Possible causes for this could be obstruction of ureters, renal damage or dehydration. Most commonly, visceral gout following dehydration is observed in newly hatched chickens after overheating or a more prolonged stay in the hatchery. Articular gout is characterized with peri articular urate deposits, especially around the joints of toes and the foot. The joints are enlarged and toes are malformed. After opening of affected joints, the peri articular tissue will be white. A number of etiological factors are related to this condition are excess protein, calcium, sodium bicarbonate, mycotoxins, and lower phosphorus levels in the feed.

Cage layer fatigue: Cage layer fatigue is a nutritional disease of chicken affecting soft long bones, bone becomes bowed and it affects the ability of the bird to stand and walk. Generally, the term rickets is used to describe the condition in young poultry and osteomalacia is often for the conditon in adult birds. Cage layer fatigue is a related condition observed in caged laying hens, usually around peak egg production, similar to osteoporosis, a condition causing brittle and week bones, as a result of reduced bone density. Rickets is caused by a deficiency or imbalance of circulating calcium, vitamin D3 and phosphorous. Cage layer fatigue is caused primarily by depletion of body stores of calcium as a result of delay in feeding high calcium feeds during high egg production or impairment in calcium absorption or bone calcification during this production stage. Excess of either calcium or phosphorous can cause rickets. Vitamin D3 plays a critical role in regulating the absorption and metabolism of calcium. Therefore, in addition to ensure that diets have an appropriate level and balance of calcium and phosphorous, they must be adequate in vitamin D. Bone mineralization is a constant process and therefore correction of dietary deficiencies or imbalances can prevent the condition. There is a higher incidence of bone calcification problems in high producing layer hens housed in cages rather than floor based housing systems.

Fatty Liver Syndrome: Fatty liver syndrome is a condition that affects only hens, primarily caged layers. It is a metabolic condition characterized by general obesity with an enlarged, fatty liver that becomes soft and easily damaged. The chief causative factor of this condition is excess calorie intake. Low protein, high energy diets, with amino acid imbalance are contributory factors to this condition in layers. Deficiency of lipotrophic factors like methionine, choline, vitamin B_{12} can also cause fatty infiltration of the liver. The hen of higher production is more susceptible to this condition. Mortality rates vary and death is often caused by internal hemorrhage due to rupture of the liver. Adjusting energy intake in the form of carbohydrate, increase the protein level in the feed, adding fat as energy source, supplemental choline, B_{12} and Vitamin E are strategies for prevention of fatty liver syndrome.

9

Water Quality Management

Learning Objectives

1. Describe the importance of water management for poultry farm
2. Discuss the importance of water quality standards of a layer farm

Water is a basic, but critical nutrient in commercial poultry feeding strategies. It is essential for all basic functions of the body system of bird. But, water management in poultry farms is the most neglected area often in commercial poultry farms. The requirement of water will be high for commercial layers, because of its high metabolic activity for its high egg production performance. In general, the water intake of commercial chicken will be twice the quantity of feed intake. For example, when the feed intake per day is 100 gram, the water intake will be 200 ml per bird per day, when the temperature is around 21°C (70°F). When the temperature increases above 35°C (95°F), water intake will be of three fold of feed intake. Moreover, poor quality water will be a source of infection for birds, leads occurrence of many diseases. Considering that, quality water should be made available to birds and the same should be available to the birds for 24 hours on *ad libitum* basis.

Water Requirement for Commercial Farms

The first important step in water quality management is checking the water quality and availability before starting a farm. It means, before starting a commercial layer farm, water quality and availability of the farm site should be checked thoroughly. The sites with poor water quality should not be selected for commercial farms. Considering the quality of water, water from deep bore wells (300-600 ft) will have stable water quality than shallow bore wells (less than 300 ft) and open wells (30-60 ft). Water quality of deep bore wells will not change much due to rainfall and sewage percolation problems. Considering that, deep bore well water is highly recommended for commercial layer farms. The next step is estimating the water requirement of a commercial layer farms on daily basis. Water requirement shall be calculated as 400 ml per day per adult layer chicken. For example, a layer farm with the batch size of 50,000 birds will require 20,000 litres of potable quality water per day for drinking purpose alone. While designing a poultry farm's civil structures, water tank volume should be designed to store at least two days water requirement of the farm. One cubic foot of volume of will hold approximately 28 litres and one cubic metre will hold 10,000 litres of water.

Table 9.1 Water quality standards for poultry farm

Sl.No.	Quality parameters	Desirable level
1.	pH	6.5-7.5
2.	Total Dissolved Solids (TDS)	Less than 1200
3.	Total Hardness	110 ppm
4.	Calcium	60 ppm
5.	Chloride	14 ppm
6.	Sodium	32 ppm
7.	Magnesium	14 ppm
8.	Nitrate	10 ppm
9.	Total bacterial count	100 cfu per ml
10.	Coli form count	Nil

Water Sanitation

Water sanitation is a process of disinfecting water through artificial means and makes it microbiologically safe for drinking purpose. The common disinfection procedure is chlorination. Chlorination is a method of adding chlorine as disinfectant to water to make the water safe for drinking. Bleaching powder chemically known as Calcium hypochlorite ($Ca (Clo)_2$) is used as chlorine source for commercial use. The quantity of bleaching powder requirement for effective disinfection is four milligram per litre of water (assuming the strength of chlorine in commercial bleaching powder is 25 per cent). Based on that, 1000 litre of water requires four gram of commercial bleaching powder. The contact time for effective chlorine disinfection is 30 minutes. The other commercially used disinfectant is hydrogen peroxide (H_2O_2) at the rate of 10 ml per 100 litre of water with the contact time of 20 minutes.

Routines in Watering Management

In deep litter system of rearing, appropriate drinkers should be used for different age group of birds. In cage system of rearing, nipple and cup drinkers will be used and height of nipple and cup should be adjusted to the height of the birds for easy watering. Number of nipples in a cage box should be adequate for effective watering. Leaky nipples should be verified on daily basis to avoid wet litter problems. Periodical checking of water quality for at least for once in three months is essential to ensure quality water supply. In case of incidence of poor water quality, disinfection of water is highly recommended as per the advice of the poultry consultants.

10

Good Husbandry Practices in Commercial Layer Farming

Learning Objectives

1. Discuss the role of Good Husbandry Practices in poultry welfare
2. Identification of critical areas for implementation of GHP in a layer farm

Definition of Good Husbandry Practices

Good Husbandry Practices (GHP) in poultry farming can be defined as a collection of package of practices to be followed in intensive or industrial poultry production systems to maximize / optimize the production and productivity without compromising the poultry welfare and food safety. However, Good Husbandry Practices is not a new set of practices of farming or production methods, but they are all refined or improved methods of our regular production activities in the production process, to achieve the overall goodness in the production to market value chain.

Benefits of GHP in Poultry Production Value Chain

Application of Good Husbandry Practices in poultry production will offer the following benefits to the poultry producer and consumer.

1. Satisfaction of producer and consumer
2. Improved marketability of the products

3. Food safety
4. Improved poultry welfare
5. Optimization of return per unit investment
6. Maximization of production and productivity

What is the Special Need for GHP in Layer Chicken Farming?

Globally, commercial production of chicken egg is characterized by intensive rearing systems. Nowadays, in all the intensive system of chicken egg production, bird welfare, food safety and pollution are becoming the issues of discussion, more than maximizing the production and productivity. Though, we get maximum production benefits in the intensive system of production, the welfare and pollution are very serious issues to be taken care of by producers (often neglected) for a holistic, humane and sustainable production environments. In this context it is essential to mention that, in industrial poultry production, capital investment is very high. Consequently, there is every possibility of providing inhumane and callous treatment towards chicken welfare considering the cost economics. In this juncture, sensitization of the stakeholders about the Good Husbandry Practices to be followed in intensive production systems to maximize the productivity with improved welfare of chicken is the need of the hour.

Poultry Welfare

As per the dictionary (Oxford / Webster) references, welfare associates with state of wellbeing, health, happiness and successful progress in life, which may rightly suits for human welfare. However, the definition of health, happiness and progress is a highly subjective term in human life, which is based on individualistic culture and priorities. Application of the same context in the area of poultry welfare may not be right and logical, since these birds (layers for egg) have specific and pre-determined purposes for their living. Based on that, it will not be logical to account the 'happiness factors' in animal welfare, because often they are not directly measurable and sometimes over ride by the production function.

Considering these background, it is reasonable to take up the definition of animal welfare proposed by World Organization for Animal Health (OIE) that is *"how well an animal is able to cope with the conditions in which it lives"*, though, this definition lacks complete acceptance from animal welfare activists. The reasons for the lack of acceptance of this definition is, it does not accounting the criticality of animal's feelings and experiences like happiness or stress factors. However, considering the perspective of Good Husbandry Practices in poultry farming, the definition of OIE will holds good to achieve the purpose. Because, as per OIE, good housing, feeding, management, disease prevention, humane handling are the mandatory factors for a holistic poultry welfare programme.

Principles of Animal Welfare

Describing and assessing the poultry welfare in intensive chicken egg production environments is complex and often debatable. In a holistic view, layer bird welfare is based on various components and can be broadly classified in to two namely effective performances of production function like egg production and other factors like happiness, freedom to lead natural way of life of its own etc. For an example, a commercial layer chicken is able to produce an average of 310-320 eggs per year in cage rearing indicates its state of well-being. However, in other way, these birds may not be at welfare state, while considering its limitations in its freedom as fowl to move in a range land. Therefore, it is important to measure each component of welfare independently and integrating the different measures to reach an overall conclusion about welfare status. However, there are some guiding principles are available based on various research works to assess the overall welfare quality and they are called as Five Freedom approach in animal welfare.

The Five Freedoms, Principles and Criteria for Good Animal Welfare

United Kingdom's farm animal welfare council proposed that, all farm animals should have five freedoms to ensure minimum welfare environments in their rearing systems and they are

1. Freedom from hunger and thirst
2. Freedom from discomfort
3. Freedom from pain, injury and disease
4. Freedom to express normal behaviour
5. Freedom from fear and distress

Table 10.1 Welfare principles and criteria in animal welfare

Welfare principles	Welfare criteria (measures)
Good feeding	Absence of prolonged hunger (feeding space)
	Absence of prolonged thirst (drinker space and quantity of water intake)
Good housing	Comfort around resting (cleanliness and litter quality)
	Thermal comfort (absence of panting and huddling)
	Ease of movement (optimal stocking density)
Good health	Absence of pain
	Absence of injuries (lameness, hock burn, foot pad dermatitis)
	Absence of disease (mortality, culled outs, production performances)
Animal Behaviour	Expression of social and other behaviours
	Good human animal relationship (Avoidance Distance Test)
	Positive emotional state (Quality Behavior Assessment-QBA)

The reason for accepting welfare definition of OIE in *Good Husbandry Practices in Poultry* is, OIE considers these five freedom approach as guiding principles for assessing the welfare of the poultry and, it also being endorsed by most of the well regulated European welfare legislations and forms the basis for OIE Terrestrial Animal Health Code. However, there is different school of thought that, it is very difficult to find out the natural behaviour of an animal in captive environments and in most cases adaptation will become an overriding factor in expression of normal behaviour, thereby problems in assessing the overall welfare quality. For an example, though *wing flapping* is normal behaviour, caged hens flaps its wings occasionally due to adaptation to captive environments. In the same way, high stocking density, limitations in scratching the ground, litter bathing, privacy for egg laying, over grown claws and leg weakness are the welfare issues in cage rearing of layer chicken.

European Welfare Quality Consortium improved this approach and suggested guiding principles and criteria for animal welfare quality as follows,

Assessment of Poultry Welfare

The scientific assessment of poultry welfare is usually a complex process, which is based on number of parameters. It includes a range of physical, physiological, behavioural or clinical indicators, and comparing these measurements among chickens that are reared in a specific and similar production system. A broad range of indicators can be used to assess stress response and immune function in an attempt to measure whether the animal is coping with its environment or not. However, interpretation of these indicators is sometimes difficult. The measures of welfare can be classified in to two as they are as follows:

Table 10.2 Resource and animal based measure of layer chicken welfare

Resource based measure for poultry	Animal based measure for poultry
Stocking density / cage space	Livability / Mortality
Feeder and drinker space	Per cent culled
No. of feeders and drinkers	Lameness, cage layer fatigue
Feed quantity	Feather loss, Hock burn
Micro environment (temperature and relative humidity	Injuries of birds
Air quality (ammonia)	Human relationship (Avoidance Distance Test)
Litter quality (moisture & freshness)	Quality Behavior Assessment

Conventionally, we used to think that optimization of resource based measure will ensure welfare, but it is not necessarily true always. Combination of these two measures of welfare will provide picture about the actual status of welfare in production environment.

In commercial chicken egg production systems, relationship between welfare and production goals is bi directionally associated (positive or negative), which is based on the specific production function and goal. It can be explained as improved methods of feeding, disease prevention (welfare measures) will ensure the welfare of poultry and lead to high production gains, in which, welfare and production is positively associated. On the other hand restricting feed intake during grower phase in commercial layer chicken will avoid the over eating and excess body weight and improvement in the uniformity. Uniformity in body weight in the specific ages, in turn improve the production function of egg production. But it may be a factor for chronic hunger, which is against welfare for these birds. In this situation production gain and welfare are negatively associated.

When production gains can be achieved by improving animal welfare, there is no need for any safe-guard mechanism to ensure poultry welfare, because of its dual benefits. But, when increased production gains, conflicts with welfare, some mechanisms are required to ensure a reasonable welfare with human touch. In these circumstances, how to ensure reasonable welfare without compromising production and productivity is a challenging situation, for which application Good Husbandry Practices will be a useful strategy. These programs may be made available in the form of law, codes of practice and voluntary welfare assurance programs etc. for effective implementation.

Welfare Issues in Commercial Poultry Production

Identification of welfare issues in intensive chicken egg production is a basic process to implement Good Husbandry Practices, and to improve the welfare. Following are the indicative list of welfare issues in intensive poultry production environment. Indicative list of welfare issues in commercial chicken egg production are as follows,

- Transport of day old chicks
- Brooder space and temperature
- Cage space and stocking density
- Micro environment in terms temperature, relative humidity, ventilation and air quality
- Feed and water quality
- Feeding and drinking space
- Feed restriction during growing
- Beak trimming, vaccination and medication
- Culled outs and mortality
- Leg problems like cage layer fatigue, growth of claws

Benefits of GHP and Welfare

Implementation Good Husbandry Practices in poultry farming offers benefits in three forms namely, improved production and consumer acceptance, wide access to markets and creation of employment. Growing awareness on welfare issues of in intensive poultry farming among the consumers of poultry products, make the people to choose products from chicken raised in good welfare conditions, which leads improved acceptance of those products without any hesitation, thereby providing satisfaction to producer and consumer.

Activity

1. Visit a commercial chicken layer farm and take an account of the resource based measure in assessing the layer welfare.
2. Compare two commercial layers farm's welfare by calculating the stocking density.

11

Commercial Chicken Layer Farm Economics

Learning Objectives

1. List out the various costs of production of chicken egg
2. Prepare of a project report for a commercial layer farm

Introduction

Understanding the economics of a commercial layer farm is the final step in the learning process about the subject of commercial chicken egg production. It will be helpful for financial planning of a commercial layer farm and to prepare a project report for availing financial assistance from banks. The cost of production in commercial layer farm implies, the cost of production per egg, which includes, feed cost, chick cost, growing expenses cost, over heads etc. However, in financial terms cost may be classified in to three namely fixed investment, fixed cost and variable cost. Understanding these costs will helpful to work out economics in a proper way.

Fixed investment: Fixed investment is the investment made by the promoter or entrepreneur to develop the poultry farm facilities, which includes, purchase of land, land preparation, construction of poultry houses, cages, equipment, facilities for water and other physical structures like office, quarters, godown and outlet if any etc. Generally, in fixed investment, 75 per cent will be provided by the financial institutions and 25 per cent will be borne by the promoter.

Fixed cost: Fixed cost is the cost which is fixed in nature and does not vary with volume of production. Cost on fixed investment in the form of interest and depreciation on building and equipment value (assumed cost) are come under fixed cost.

Variable cost: Variable cost is the cost, which is variable in nature and, will vary with the volume of production. Chick, feed, labour, electricity, litter, health management inputs, other utilities and over heads will come under the variable cost.

Total cost: Total cost is the cost which includes both fixed and variable cost.

Critical Factors of Farm Economics

Economics of commercial layer chicken farm depends principally on three factors for its profitability and they are, volume of production, cost of inputs and performance of the farm and sale price of the output. Volume of production favours the cost economics and it is called as *economy of scale* and, cost of production per unit will be minimal in large scale. Lower the cost of inputs favours the economics positively. Technical performance indicates the realized performance of a layer farm in their economic traits. High sale price of the output favours the economics positively.

Model Economics of a Commercial Chicken Layer Farm

Assumptions

1. Independent commercial layer farm
2. Type of farm : All layer farm - All layer farm refers the rearing system of only layers through direct purchase of ready to lay pullets (16-18 weeks) for egg production
3. Farm size : One lakh layer bird capacity with batch size of 25,000 layers
4. System of housing and rearing : Open sided elevated cage system of rearing

Table 11.1 Technical standards of commercial layer chicken farm

1.	Cost of land - Nil	:	Farmer's own land
2.	Cost of land preparation	:	Rs. 5,00,000/-
3.	Cost of construction elevated cage layer houses	:	Rs. 400 per bird
4.	Floor space requirement	:	0.6 sq.ft per bird
7.	Cost of point of lay pullet	:	Rs.200 per bird
8.	Mean egg production period	:	52 weeks
9.	Mean saleable egg number per bird	:	300
10.	Feed intake from 20-72 weeks	:	42 kg
11.	Cost of feed per kg	:	Rs. 20
12.	Cost of health care and other charges	:	Rs.5 per bird
13.	Mortality during laying period	:	10 per cent
14.	Mean body weight at culling at 72 weeks	:	1.6 kg
15.	Mean culling age	:	72 weeks
16.	Sale price of culled birds	:	Rs.50 per kg
17.	Cost of labour per month		Rs. 96,000
18.	Depreciation on building and equipment	:	10 per cent
19.	Interest on investment	:	15 per cent

Table 11.2 Economics of chicken layer farm (one year production)

Sl.No.	Items	:	Amount (Rs.) (in lakhs)
A)	**Fixed investment**		
A1	Cost of Land	:	Nil
A2	Cost of preparation of land for layer farm	:	5.00
A3	Cost of construction layer houses with accessories (Rs.400 x 1 lakh birds)	:	400
A4	Total	:	405
B)	**Fixed cost**		
B1	Interest on investment @ 15 % per annum [Rs.405 lakhs x 0.15]	:	60.75
B2	Depreciation on building @ 10 % per annum [400 lakhs x 0.10]	:	40.00
B3	Total	:	110.75
C)	**Variable cost**		
C1	Cost of point of lay pullets [One lakh x Rs.200 per chick]	:	200.00
C2	Cost of feed [One lakh x 42 kg x Rs.20]	:	840.00
C3	Cost of medicines, vaccines and other utilities [One lakh x Rs. 10 per bird]	:	10.00
C4	Cost of labour - 12 labours per day @ Rs.7500 per month [96,000 per month x 12 months]	:	10.80
C5	Total	:	1060.80
D)	**Total cost [B plus C]**	:	**1171.55**
E)	**Income / Receipts**	:	
E1	Sale of market saleable eggs [One lakh x 300 eggs per bird x Rs.4.00 per egg]	:	1200.00
E2	By sale of culled birds at the age of 72 weeks [90,000 birds x 1.6 kg x Rs.50 per kg)	:	72.00
E2	Gross income	:	1272.00
E3	Net income [E minus D]	:	100.45
E4	Net income per layer [E3 divide by (1,00,000 birds)]	:	Rs.100.45

Annexure

Metric Equivalents

1 meter	100 cm
1 meter	3.28 ft
1 feet	12 inches
1 inch	2.54 centimeters
Area	
One hectare	2.47 acres
One acre	4046 square meters
One square meter	10.76 square feet
One square feet	144 square inches
One square feet	929 square cms.
Area of circle	Pi r square
Weight	
One kilo gram	2.20 pounds
One pound	0.45 kilograms
One ton	1000 kilograms
Temperature	
Centigrade to Fahrenheit	°C x 1.8 plus 32
Fahrenheit to Centigrade	(°F minus 32) / by 1.8
Concentration	
One mg per kg	One ppm
One micro gram per gram	One ppm
One micro gram per kg	One ppb
Volume	
One cubic foot	28 litres (approx.)
One cubic meter	1000 litres
Milligram per litre	1 ppm
Gram per litre	1000 ppm

Index

A

Abnormalities 58

Accessibility 11

Agricultural and processed food products export development authority (APEDA) 72

Alfalfa leaf meal 45

Alimentary canal 34

Anas platyrhynchos 1

Animal welfare 94, 95

Anser anser 1

Antibiotic supplements 46

Arachidonic acid 70

Attenuated vaccine 77

Avian cholera 86

Avian encephalomyelitis 84, 85

Avian influenza 82, 83

Avian pox 83

B

Bacterial diseases 85

Barred plymouth rock 5

Basic function 34

Beak trimming 27, 28, 97

Bio-security 73, 74, 75

Blue wing disease 85

Bone meal 44

Brooder house 20, 22

Brooder pneumonia 87

Brooding equipment 23

Brooding management 23

Brooding period 23

Brown egg shell layer 6

Building plan 22

Built up area 11, 12, 20

Bureau of Indian standards 39, 40, 67

Butylated hydroxyl anizole 46

Butylated hydroxy toluene 42, 46

C

Cage brooding 24

Cage design and arrangement 16

Cage house 12, 17

Cage layer fatigue 89

Cage rearing 15

Cage specifications 17

Calorie Protein Ratio 37
Chronic respiratory disease 86
Coccidiosis 88
Coefficient of variation 28
Colibacillosis 86
Commercial chicken egg production 1, 70
Commercial layer houses 13
Common vaccination methods 77
Conceptual bio-security 73
Construction coefficient 11, 12
Cooked eggs 66
Coturnix coturnix japonica 1
Critical factors 3, 100
Culling 29

D

Dead bird disposal 75
Deep litter house 12
Digestive System 34
Disinfection 75, 76
Docosahexanoic acid 70
Double-yolk eggs 58
Dromaius novaehollandiae 1
Dry cleaning 75
Duck 1

E

Egg candling 59
Egg collection 9, 15
Egg drop syndrome 85
Egg formation 57, 58
Egg number 5, 6, 22, 28, 29, 30, 100
Egg production 1, 3, 5, 28, 31, 32, 105
Egg quality standards 67
Elevated cage houses 17, 22
Elevated open sided cage house 18
Emu 1
Energy sources 40
Enriched cages for chicken layers 19
Environmentally controlled cage houses 18
Erucic acid 43
Essential amino acids 37
Extensive system of rearing 10
Exterior quality factors 61

F

Farm area 11
Farm economics 99, 100
Farm site 11
Fatty liver syndrome 89
Feed additives 45
Feed enzymes 46
Feeding and watering space allowance 14
Feed ingredients 39
Feed milling process 48
Feed restriction 26, 28, 50, 51
Female reproductive tract 57
Fixed investment 99, 101
Flock uniformity 28
Floor brooding 23
Floor level cage houses 17, 22
Formaldehyde gas fumigation 76
Fowl pox 79, 83
Full fat soybean 42
Fungal diseases 87

G

Gallus domesticus 1
Genetically modified 8
Genetic stocks 5
Good husbandry practices 3, 16, 93-97
Goose 1

Groundnut meal 43, 45
Grower house 20, 22
Grower management 26
Guinea fowl 1
Gumboro disease 82

H

Haugh unit score (HUS) 64
Hydrolyzed poultry feather meal 44

I

Identification and selection 11
Immunoglobulin yolk 70
Industrial layer chicken breeding programme 7
Infectious bronchitis 79, 82
Infectious bursal disease 79, 82
Infectious coryza 85
Infectious laryngotracheitis 83
Intensive system of rearing 2, 10
Interior quality factors 61
Intramuscular injection 78

J

Japanese quail 1
Japanese vent sexing 21

K

Killed vaccine 77

L

Laboratory examination 48
Large intestine 35
Layer chicken breeding 6
Layer chicken management 21
Layer house 20, 22
Lighting management 29
Live viral vaccines programmes 78
Locality 11
Lymphoid leucosis 84

M

Maize gluten 40
Marek's disease 79, 84
Mash type 49
Meat by products meal 43
Meleagris gallopavo 1
Metabolic diseases 89
Metabolisable energy 37
Millets 40
Model economics 100
Mycoplasma synoviae infection 87

N

Nasal route 77
National egg coordination committee 71
National research council 39
Natural incubation 21, 23
Newcastle disease 79, 81
New Hampshire 2, 5
Non starch polysaccharide 46
Numida meleagris 1
Nutraceutical properties 70
Nutrient feed additives 46
Nutrient requirements 38
Nutritive value 59

O

Ocular route 77
Oesophagus 34
Open sided battery cage layer houses 19
Open sided house 12
Operational bio-security 74
Orientation 13
Ornithology 1
Ostrich 1
Oyster shell 45

P

Pancreas 36
Peak egg production 29
Pellet type 49
Phase feeding 50
Physical structure 55
Poly unsaturated fatty acids 43, 70
Poultry carcass disposal 80
Poultry diseases 81
Poultry farming 33
Poultry feeding standards 39
Poultry feed milling 47
Poultry houses 12
Poultry welfare 94, 96
Power supply 11
Pre lay diets 50
Protein sources 42
Pullorum disease 87
Pure line breeding 8

Q

Qualitative feed restriction 26
Quality control 47
Quality testing 48

R

Rapeseed meal 43
Rearing in chicken egg production 9
Red jungle fowl 1
Rendering 80
Restricted feeding 50, 51
Rhode island red 2, 5

S

Safflower meal 43
Semi intensive system of rearing 10
Sexual maturity 6, 7, 22, 27, 28, 30, 32, 50, 84
Shell colour and texture 62
Sialic acid 70
Single comb white leghorn 5
Small intestine 35
Soy bean meal 42, 49
Structural bio-security 73
Struthio camelus 1
Subcutaneous injection 78
Summer management 51

T

Topography 11
Turkey 1, 44

U

USDA egg quality standards 65, 66

V

Vaccine forms 77
Variable cost 100, 101
Viral diseases 81
Visceral leucosis 84

W

Water availability 11
Watering management 92
Water quality management 91
Water sanitation 92
Welfare principles 95
Wet cleaning 75
Wheat bran 41, 44
White egg shell layer 6
World organization for animal health 94
Worm infection 88

Y

Yolk 56
Yolkless eggs 58

www.ingramcontent.com/pod-product-compliance
Ingram Content Group UK Ltd.
Pitfield, Milton Keynes, MK11 3LW, UK
UKHW020047200726
13845UKWH00044B/21

9 789390 371983